PENGUIN BOOKS

ANATOMIES

Hugh Aldersey-Williams studied natural sciences at Cambridge. He is the author of several books exploring science, design and architecture and has curated exhibitions at the Victoria and Albert Museum and the Wellcome Collection. His previous book, *Periodic Tales: The Curious Lives of the Elements*, was a *Sunday Times* bestseller, has been published in many languages around the world and is available in Penguin. He lives in Norfolk with his wife and son.

Anatomies

*The Human Body, Its Parts
and the Stories They Tell*

HUGH ALDERSEY-WILLIAMS

PENGUIN BOOKS

PENGUIN BOOKS

Published by the Penguin Group
Penguin Books Ltd, 80 Strand, London WC2R ORL, England
Penguin Group (USA) Inc., 375 Hudson Street, New York, New York 10014, USA
Penguin Group (Canada), 90 Eglinton Avenue East, Suite 700, Toronto, Ontario, Canada M4P 2Y3
(a division of Pearson Penguin Canada Inc.)
Penguin Ireland, 25 St Stephen's Green, Dublin 2, Ireland
(a division of Penguin Books Ltd)
Penguin Group (Australia), 707 Collins Street, Melbourne, Victoria 3008, Australia
(a division of Pearson Australia Group Pty Ltd)
Penguin Books India Pvt Ltd, 11 Community Centre,
Panchsheel Park, New Delhi – 110 017, India
Penguin Group (NZ), 67 Apollo Drive, Rosedale, Auckland 0632, New Zealand
(a division of Pearson New Zealand Ltd)
Penguin Books (South Africa) (Pty) Ltd, Block D, Rosebank Office Park,
181 Jan Smuts Avenue, Parktown North, Gauteng 2193, South Africa

Penguin Books Ltd, Registered Offices: 80 Strand, London WC2R ORL, England

www.penguin.com

First published by Viking 2013
Published in Penguin Books 2014
001

Typeset by Palimpsest Book Production Limited, Falkirk, Stirlingshire
Printed in Great Britain by Clays Ltd, St Ives plc

ISBN: 978–0–670–92072–3

www.greenpenguin.co.uk

MIX
Paper from
responsible sources
FSC™ C018179

Penguin Books is committed to a sustainable
future for our business, our readers and our planet.
This book is made from Forest Stewardship
Council™ certified paper.

To Moira

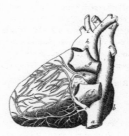

Contents

List of Illustrations

Acknowledgements

Our attitudes towards the human body are so confused and conflicted that on many occasions during the research for this book I found my access barred to the things that I felt I needed to see and experience – barred ostensibly by regulation, but in fact by timid gatekeepers who did not want the trouble of opening their resources to an outsider's gaze. I am all the more grateful, then, to those few who were prepared, in the face of these regrettable restrictions, to grant me a privileged view into what is in fact our own corporeal world. I am grateful above all to Sarah Simblet of the Ruskin School of Drawing and Fine Art, who allowed me to join her classes in drawing from anatomy, and to John Morris, the professor of human anatomy at the University of Oxford, in whose department this uniquely informative activity takes place.

Ken Arnold provided the introduction to Sarah without which this book would have had nowhere to begin. Once again I am greatly indebted to him and his colleagues at the Wellcome Collection, James Peto, Lisa Jamieson, Rosie Tooby and Elayne Hodgson, for their assistance and expertise. In 2009, they were kind enough to invite me to curate an exhibition called 'Identity: Eight Rooms, Nine Lives'. I have borrowed greedily from some of the lives that we presented there. I am immensely grateful to April Ashley, who allowed us to tell the remarkable story of her gender reassignment for that exhibition, which I have retold in brief here. I am grateful, also, to Ruth Garde, who delved into the phrenological literature and came up with many riches, a few of which I also draw upon, and to the various neuroscientists whose fMRI images of the brain featured in that exhibition. A small part of my chapter dealing with the brain is adapted from an essay I wrote for the catalogue of that exhibition, *Identity and Identification* (London: Black Dog Publishing, 2009).

This is the first book that I have written about the life sciences, and one of the chief pleasures associated with it has been my discovery of the Wellcome Library. Here I was imaginatively guided by William Schupbach, Simon Chaplin, Ross Macfarlane, Christopher Hilton and Lesley Hall. Diana Wood at the library of the Fitzwilliam Museum and the staff of the Cambridge University Library also provided assistance.

I would also like to thank Fay Bound Alberti, Sam Alberti and his colleagues Carina Phillips, Tony Lander and Martyn Cooke at the Royal College of Surgeons, Santiago Alvarez, Vittorio and Enrica Norzi, Andrea Sella, Erik Spiekermann, Luba Vikhanski, Barbora Koláčková and Jana Vokacova, who responded enthusiastically to my request for body idioms in languages other than English, Derek Batty, Sarah-Jayne Blakemore, Barry Bogin, Serena Box, Vicki Bruce, Edwin Buijsen at the Mauritshuis, Deborah Bull and Molly Rosenberg at the Royal Opera House, Chris Burgoyne, Gemma Calvert, Emily Campbell, Emma Chambers, Alex Clarke, Jody Cundy, Chris Furber and Iga Kowalska-Owen of the British Paralympic cycling team, Andrew Douds, Alan Eaton, William Edwards at the Gordon Museum, Pascal Ennaert at the Groeninge Museum, Mattie Faint, Chris Frith, David Gault, Roderick Gordon, Michael John Gorman and Brigid Lanigan at the Science Gallery, Dublin, Daniel Green, Gary Green and Sam Johnson at the York Neuroimaging Centre, Aubrey de Grey, Annabel Huxley, Karen Ingham, Jim Kennedy, Tobie Kerridge, Vivienne Lo, Natasha McEnroe, James Neuberger, Helen O'Connell, Deborah Padfield, James Partridge, David Perrett, Wolfgang Pirsig, Emma Redding and her colleagues Mary Ann Hushlak, Sarah Chin and Luke Pell at the Laban Centre, Keith Roberts, Laura Bowater, Hope Gangata and David Heylings at the University of East Anglia, Nichola Rumsey, Volker Scheid, Don Shelton, Jim Smith, Charles Spence, Lindsay and Justin Stead, Viren Swami, Julian Vincent, Crawford White, Fiona Wollocombe, Duncan X and Blue at Into You. I confess to stealing the idea for the illustration on the dedication page from Ruth Richardson.

I happily thank my agent Antony Topping, my editor Will

Hammond, copy editor David Watson and my wife Moira and son Sam, who have once again put up with me as I battled to learn something of a topic about which, like most of us, I knew and still know so little.

Hugh Aldersey-Williams
Norfolk, July 2012

These little Limbs,
These Eyes and Hands which here I find,
These rosy Cheeks wherewith my Life begins;
Where have ye been? Behind
What curtain were ye from me hid so long?
Where was, in what Abyss, my new-made tongue?

From 'The Salutation', Thomas Traherne, 1637–74

Introduction

There comes a moment in your life when you realize that you are probably not, after all, going to be the first person to live for ever.

This realization is normal, or at least I assume so, but it is quite contrary to your rightful expectation. It is a shock.

Let's be clear. Your mind has no problem with the idea of living for ever. What's wrong with just carrying on as it is doing now? It can see no reason not to. No. Your body's the problem. It begins to work less well. And it starts going on about itself, sending you ever more frequent messages, nagging, needy messages: What about me? Is nobody listening to me? Stop that, it hurts. Or: 'I need to go to the loo.' 'What, now?' your mind sleepily responds. 'It's three in the morning.' 'Yes, now.'

At school, I was obliged to give up biology at the age of thirteen, even though I was beginning to veer towards the sciences. One lesson a fortnight became no lesson at all. It now seems to me astonishing that this should be allowed to happen, not only because it was by then already clear that biology was the scientific speciality which offered the greatest scope for discovery, but also because we are each the owner-operator of our own human body, and this surely would have been the time to learn something about it. Instead, I was left in charge of a complex biological organism about which I knew next to nothing and yet, if I was lucky, I would be instructing and inhabiting for another seventy years or so.

One consequence of this educational omission and my own intellectual laziness is that I have no answer to the bathroom-at-three-in-the-morning problem. I have no idea how my bladder works, or why it seems to work differently now from the way it worked when I was younger. And the chances are that you haven't either.

I have only a vague picture of some sort of watertight balloon that fills and empties itself, located somewhere in my abdomen. For more precise information, I find I must consult an undergraduate text-book. The book is like a paving slab, and it's full of colourful but artless drawings. I look up 'bladder' in the index. It is not there. I see I'm going to be forced to translate my simple enquiry into an alien jargon. I consider the matter briefly, and then turn to the Us to find an entry for 'urinary system'.

The bladder, I learn, is an elasticated bag made of thin layers of muscle. It is lined internally with mucus to make it watertight. When full, it distends to the size and shape of a large avocado and contains about a pint of urine (or a litre, according to another textbook, nearly twice as much). An accompanying X-ray image, referred to – as if I'm meant to know the meaning of the word – as a pyelogram, shows the arrangement of the urinary system within the body, highlighted by a contrast agent that has been injected into the human subject. I see a bulbous reservoir cradled by the pelvic bones at the base of the spine. Running out of it are the two thin lines of the ureters, which hug the spine on their way up to the kidneys, where each branches into two, and then five, and then many more thinner tubes that ter-minate in the depths of each kidney, somewhere about level with the bottom rib of the rib cage. The image is rather beautiful, like two long-stemmed irises in a bulb vase.

The ureters are muscular tubes that squeeze the urine produced by the kidneys down into the bladder. As the bladder reaches capacity, stretch receptors in the muscle wall are stimulated, sending signals to your brain that you construe to mean that you should get up and relieve yourself.

Well, not exactly. In fact, the system is smarter than that. The bladder sends out its first signals simply to test your readiness. On this occasion, your brain responds to the news by sending a message back to the bladder that causes its muscles to contract a little, increas-ing the pressure of fluid within. The purpose of this is to gauge whether another set of muscles, the ones that allow the bladder to drain when they are relaxed, will hold for a bit longer. The brain is stalling, asking the bladder in effect: 'Do you really mean it?' When the bladder sends back signals admitting that it was just a bluff, your brain responds with an instruction to the muscles of the bladder wall to relax again and allow more urine to accumulate. All this happens in your sleep, and saves your being awoken until you really need to be. It's like the snooze button on an alarm clock.

The textbooks gloss over the deterioration of this admirable system in middle age. I try to reason it out. Maybe your bladder contracts, meaning that it must be emptied more often. Or maybe it expands, triggering the stretch receptors sooner. Maybe the stretch receptors themselves become more twitchy. Maybe the neural telegraph between your brain and your bladder goes a bit haywire and starts sending the wrong messages. Maybe your ageing brain just panics and thinks better safe than sorry. There are so many possibilities. I run my theories by a friend who is a hospital consultant. 'I've been trying to look it up myself,' he tells me after a while, but the process has left him as baffled as I am. Finally, he passes my questions to a colleague who is a urologist. In fact, I'm told, you simply produce more urine in your sleep as you get older. It's an inconvenient truth to say the least.

It seems absurd that finding an explanation of this most banal of bodily functions has required so much expert consultation. But I have more awkward questions, too. Is the bladder just a 'bag' or is it something more? Is it an organ? What qualifies as an organ? Where do organs stop and start? Student doctors often buy themselves a plastic skeleton and a plastic model of the body in which brightly painted organs hook neatly into place alongside one another. Is this really what the body is like? Or are the organs perhaps cultural inventions, better regarded as repositories for various ideas that we have formed about life than as discrete entities in biological reality? Does it even make sense to talk about the body in parts? For whom does this make sense? And if so, is the human body merely the sum of those parts, or is it something more? For Aristotle was indeed thinking about the human body when he coined this now overworked phrase – 'more than the sum of its parts' – in his *Metaphysics*. And if the human body *is* more than the sum of its parts, then what is this 'more'?

★

Anatomies is my attempt to make amends for my missing education in human biology and to find answers to these questions. Like most of us, I know shamefully little about how my body actually works – and occasionally doesn't. Those who do know – our doctors – seem keen to keep the knowledge to themselves, guarding their professional status with their long words, their simplified attempts at explanation, and those famously illegible prescriptions.

It is obvious that the human body is a difficult subject. Perhaps we are too close to it. The human body is routinely described as a marvel of nature, but it is surely the marvel of nature we least stop to observe. When all is well, we simply ignore it. I suppose this is as it should be – no other animal spends time pondering its wellbeing, after all. But for us, ignorance is not bliss. We are frequently ashamed of our bodies and embarrassed by them.

At the same time, we are bombarded with images of the human body. These are invariably presented as more perfect versions of ourselves. They look better (supermodels) and perform better (action heroes), but essentially they are doing things we do. These proxies remind us that our body, too, is out there in the world. It is through our body that we sense the world and must interact with it. And it is through our body that we are seen and recognized for who we are.

Yet still our bodies trouble us. We disguise them with clothes. We distract attention from them with accessories, a hairstyle, a gait, a repertoire of gestures, a voice, a way with words, to the extent that these become the greater part of our personal identity. Helped by modern medical techniques, we are becoming bolder with these manipulations. From Brain Gym to boob job, we now seek to transform our minds, personalities, faces and bodies. And in truth we have always altered ourselves in both psychological and physical ways. This is just the latest chapter in a continuing story. The idea of the body as a canvas is not new. It's just that more people than ever are starting to paint.

Then there is our attitude to medicine, the science of maintaining and restoring bodily health. In most sciences, there is some respect for history. Scientists may not refer much to the past in their field, or even know its key figures and dates, but they readily concede that the discoveries of today are built on those of the past, and that we see further because we stand on the shoulders of giants. The history of human biology and medicine, on the other hand, is easily mocked. We laugh at the once widespread belief that you can deduce somebody's character from the bumps on their head. We laugh at the hopeless cures and painful procedures of the past – the idea, for example, that field-mouse pie was an effective remedy for whooping cough. We laugh because laughter is a response to fear, and we fear for the fragile workings and the ultimate disposability of the human body – of our own human body.

Meanwhile, science is taking us in a new direction – deeper. Slowly, we are getting used to the idea that we will learn most about our bodies by zooming in on them – to examine the cells, the genes, the DNA, the proteins and other biological molecules that make us the way we are. We are given to understand that the key to the body's functions and its malfunctions – our diseases – lies in the codes and sequences that determine the character of these minute components, and in the chemical reactions that occur and the electrical signals that pass between them.

This is exciting and specialized work. For a privileged few, it offers a newly informed view of the body. But it is a very partial view. It may be that the human person can be described as a string of letters or numbers based on these new methods of investigation, and that such a description is useful in some kinds of research. But this is not the description that interests me. This is not the description we have lived by these past tens of thousands of years as a species. This is not the image we have of ourselves. Knowing that each human body has a set of chromosomes called a genome containing more than 20,000 genes, each of those genes being described by a given sequence of DNA, and that all of these genes are present in every cell of that body is important, but it does not supersede the older knowledge that the body contains a heart, two eyes, 206 bones and a navel. It merely augments it. In its technical detail, it is the sort of description that for many of us seems to miss the point. It doesn't tell us about ourselves in the round.

'Know thyself', famously ran the inscription at the temple of the oracle at Delphi in Ancient Greece. Yet, for all our scientific prowess, we seem to know ourselves, and above all our physical selves, less and less. Perhaps it's even the case that the quest for scientific understanding of the body becomes a substitute for actual bodily experience: a recent survey of students at an American university found the highest levels of virginity among students studying biology and other sciences (the lowest levels were among those studying art and anthropology).

Something of this kind of displacement has certainly happened in medical schools. The emphasis now is on the detail, not the whole. The sense of the wholeness of the human body has been diminished by the rise of specialization, which insists that the body is regarded in terms not just of parts but of isolated parts. The need to grasp the basics of genetics, molecular biology, pharmacology, epidemiology and public health has all but forced out the teaching of human anatomy, which was central to medical education for hundreds if not thousands of years. In 1900, a medical student might have attended 500 hours of gross anatomy – the anatomy of the whole body. Today, the figure might be one-third of that. And increasingly, that anatomy is seen not in the literal flesh but as a digital image on a screen.

In short, the body is taken as read. It is assumed not only that the medical profession know all they need to know about its general arrangement and function, but that, at our much lower level of understanding, the rest of us do too. I am no medic and naturally try to avoid hospitals. Before embarking on this project, I had never seen an opened body. It's almost as if somebody has figured it's better that way. It's better that we don't know too much. Then we won't question our doctors. Then we won't worry about what actually happens to us when we fall ill and when we die.

Still, chin up. Unique among species, we are blessed as well as cursed with this awareness of our selves and our bodies. Shouldn't we use this critical distance to come to a more informed view, indeed to come to some kind of reconciliation with our mortal flesh?

Anatomies is a personal attempt to do this. As you'd expect for a subject with such a rich cultural history as the body, it draws not only on past and present perspectives in medical science, but also on views of the body and its parts taken by philosophers, writers and artists. The body is not just a thing, whether that's the object on the anatomist's table or the subject in a life-drawing class. It is animate. So I will look also at the body in action – the body that moves and performs, and expresses thought and emotion. This, as much as our genes, makes us what and who we are. But don't worry. I will spare you descriptions of my own body's shortcomings. To misquote Montaigne: 'I myself am *not* the subject of my book.'

Anatomies is arranged in chapters based on significant body parts. This provides a familiar structure, although it will quickly be seen that the content of these chapters ranges well beyond the bounds of those parts. We have an idea about all of the parts I have singled out, whether they are internal organs or visible features of our bodies, an idea that owes relatively little to modern science or medicine. Instead, it is shaped by our culture, which has bestowed symbolism and meaning on the parts of the body through long and intimate familiarity with them. To rediscover these meanings, we need to touch and feel, look and listen to the body that we think is so familiar, and which we have preferred to consider in the abstract.

One of the most powerful of these associations, for example, is the idea that the heart is the seat of love. 'Come bring your sampler, and with art / Draw in't a wounded heart', the English poet Robert Herrick wrote four centuries ago in his great poem of unrequited love. But is this meaningful now? It surely is to the shops which see Valentine's Day spending of more than £2 billion in Britain alone. It is not just as the visual icon on a million cards that the heart lives in our culture. Its pulsing rhythm may underlie the pleasures of the iambic metre in poetry and the beat of rock music.

The eye at the moment of death has long been said to retain the image of the last thing it saw. Has this myth been dispelled? Only just, maybe. In 1888, the Metropolitan Police in London photographed the eyes of Mary Jane Kelly, the last presumed victim of Jack the Ripper, in the desperate hope that they might hold the residual image of her killer.

Such beliefs reflect early attempts to understand and come to terms with our bodies. Often, modern medicine is influenced by these ideas more than it cares to admit. Take blood. Old taboos still echo through the questionnaires one must complete in order to give blood, with their strange hints at tribal purity. Our feelings about organ donation, too, are coloured by deep cultural prejudice. If donors or their kin place restrictions on the organs that may be harvested, they are most likely to be on the heart and the eyes, based on the belief that the heart is the essential core of the person and the eyes are the window into the soul.

The arts can tell us things about our body that medicine and biology do not. The head is an important part, so much so that it can stand for the whole body, as we see in the sculptor's bust or your own passport photograph. But what happens when the nose alone stands for the head? In Nikolai Gogol's short story 'The Nose', a man's nose detaches itself from his face and goes off on its own around St Petersburg, pursued by its nasally challenged owner. Importantly for the satirical bite of the tale, the nose takes on the man's social pretensions. The story raises questions about the way in which certain parts of the body constitute our personal identity and others do not. But most of all it reminds us that the body and its parts are funny, if not ridiculous – or at least our constant self-awareness makes them so.

Separated from their bodies, organs and parts sometimes multiply in alarming ways, gaining strange powers as they do so. In *Gargantua and Pantagruel*, Rabelais imagines a wall of vulvas protecting the city of Paris. 'I have noticed that, in this town, the thingummybobs of women are cheaper than stone,' observes Pantagruel's companion Panurge. 'You should build walls of them, arranging them with good architectural symmetry, putting the biggest ones in the front ranks, then sloping them back upwards like the spine of a donkey, making ranks of the medium ones next and finally of the smallest.' A portrait of Queen Elizabeth I painted towards the end of her reign around 1600 shows her wearing a dress covered with appliqué eyes and ears, emblematic of the all-knowing state of which she, of course, was the head. The artist Marcus Harvey caused uproar when he created a vast painting of the child-murderer Myra Hindley using children's handprints as individual pixels. The work adapted a photograph of

Hindley much reproduced in the newspapers at the time of her trial. Was there evil in that face? Is there good in a child's hand? What did it mean to bring the two together?

This book is about our bodies, their parts, and their multiple meanings. It is also about where we draw the limits of the body, and how we are always seeking to extend those limits, never more so than right now. For 'extend', I should perhaps write 'redraw', for although we like to think of ourselves as constantly extending the human frontier, the fact is that from time to time we choose tactical withdrawal. We draw the limits closer in, not further out. We think we like the idea of being all-capable, but in fact we'd rather not test our capacity for pain, or even make much use of our senses of smell and touch, for example. We think we'd like to live longer – or is it just that we'd prefer to avoid dying? We dream of escaping our bodies and existing in transformed or dematerialized ways. We may think these dreams are the product of recent or promised advances in biomedical technology. But in fact they are the timeless product of our imagination.

Overarching this part-by-part progress, then, is another idea: the idea of the body as geography, as territory to be discovered, explored and conquered. This powerful metaphor is found throughout human culture, from the plays of Shakespeare to the 1966 film *Fantastic Voyage*, in which miniaturized humans journey through a man's body in a quest to save his life. It also seems to reflect how science has proceeded, claiming new-found lands, dividing them into parts, proclaiming dominion over them on behalf of specialist new disciplines. It is, I might add, a very male approach, never more so than when it is the female body that is being explored.

At one point in my research, I noted something peculiar about my reading list. I was being curiously drawn to books set on islands – *Robinson Crusoe* for its all-important human footprint, *Gulliver's Travels* for the changes in human scale it imposes, *Typee* for the tattoos and the cannibals, *The Island of Dr Moreau* for the vivisection and the human-animal hybrids. Why was this? Islands present isolated populations. Here, humans are almost a subspecies of *Homo sapiens*, ripe for the kind of anthropological scrutiny that might seem impertinent among the home population. Islands are places where, for a time, you can observe and control a community as if it is part of an experiment. But the situation cannot be sustained. Eventually, the hero escapes to tell his improbable tale (or not, in the case of Dr Moreau's visitor, who pretends amnesia because what he has seen is so incredible). As John Donne famously reminds us in his *Meditations*: 'No man is an island, entire of itself; every man is a piece of the continent, a part of the main.'

These fictional island laboratories are places for the exploration not just of human nature writ large, but also of the identity of the individual. The body may be seen as territory with parts that have been more or less thoroughly explored, but somewhere within that territory, we are convinced, is a special place, the seat of the soul, as we once said, or of the self, as we might say today. In medieval times, a person's heart was often preserved or buried separately from the rest of the body because it was the part thought to be most closely associated with the soul. During the Renaissance, a more sophisticated notion took over. The soul was to be found in the divine proportions of the human body, which was the answering microcosm to the macrocosm of the ordered universe. The ideal bodies and the anatomies of this period, from Leonardo da Vinci's Vitruvian man to Rembrandt's paintings of dissections, reflect this belief. With the progress of science, however, the urge to find a focal point was soon reasserted. Attention settled on the head, as physiognomists sought their answers to the problem of the self in human expression and phrenologists in the bumps of the skull. Today, we look at magnetic resonance imaging scans of the brain and believe these bring us closer to knowing our selves. It seems that only a visual image will give us the reassurance we seek.

This need to see the self is strong because we live in a society that prizes human individualism, and also because we sense that the self is susceptible to manipulation in ways that it never has been before. We are aware that our personal identity may be altered – and perhaps improved – by consciously undertaken extension. Such extension may be psychological (self-help books), physical (cosmetic surgery), chemical (mind-altering drugs) or technological (virtual environments). At the moment, these possibilities are perhaps only crudely being tested. However, it seems certain that in future it will be increasingly easy, and probably increasingly acceptable, to manipulate both the external appearance of the body and our genetic make-up, and that this may disrupt what one bioethicist calls the former 'naturalness of the self'.

These are exciting and troubling times for the human body. We seem both excessively aware of it and yet at the same time profoundly dissatisfied with it. The biological sciences promise many things about the way we will live in future. But, however beautiful we are, however super-capable we become, however long we live, we still must inhabit our bodies. Perhaps, by recognizing the human body as a site of *continual* invention, we may overcome the distortions of the present moment.

Finally, one barrier to a broader comprehension of the body is the profusion of Greek and Latin names – names that those in the medical professions were themselves once put to great labour to learn. There is an argument that these provide a universal language much like the Mass sung in Latin, but I am not convinced. So I have tried to minimize my use of these words, many of which were baffling to me as I set out. I won't use 'anterior' where 'front' will do, or 'femur' for 'thigh bone'. It seems wrong that the parts of our own bodies should be described in a vocabulary that is alien to us.

Now, if you'll excuse me, I have to go and pee.

Prologue: The Anatomy Lesson

Who, I am wondering, is the subject of this painting?

I am at the Mauritshuis, one of the world's great collections of Dutch art, housed in a perfect little palace on the lakeside in the centre of The Hague. I have just seen Vermeer's *Girl with a Pearl Earring*. Its sheer beauty produces a choke of emotion. Now, two rooms away, I am standing in front of Rembrandt's painting known as *The Anatomy Lesson of Doctor Tulp*.

It is his breakthrough painting. Rembrandt arrived in Amsterdam in 1631 aged twenty-five looking for portrait work. He found it almost immediately when Nicolaes Tulp, the praelector, or public lecturer, of the Amsterdam guild of surgeons, asked the young artist to paint him with his fellow guild members. The job must have surpassed Rembrandt's hopes for it offered a fantastic challenge: to paint not just one man but many men, to find a way to communicate the individuality of each of them, and yet also to conform to the expectations of the seventeenth-century equivalent of a team photograph. And would there, Rembrandt must have mused as he accepted the work, be scope also to tell a more universal story?

It is a massive canvas. It shows a group of seven men, almost life-size, attending closely to Doctor Tulp, who sits in an armchair, enthroned and slightly elevated, demonstrating a point of detail in human anatomy. Yet it may not be Doctor Tulp who is the subject, as the earringed girl is so obviously the subject of the Vermeer. The painting's title only came later. It is, as it was meant to be, a genre painting of a group of professional achievers. The identities are known of the other men depicted. They too are surgeons. They may appear agog to learn, but Tulp's audience in the painting are his equally proficient peers. He has no anatomy lesson for them. So perhaps all the surgeons together are the subject. It was they who paid for the painting, and it was immediately hung on the wall in their guild.

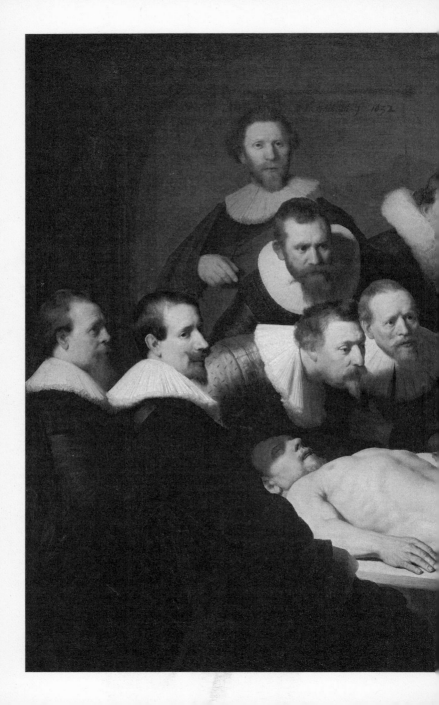

But I don't think these fellows with their florid cheeks and extravagant ruffs are truly the subject either. For us, and for Rembrandt, the true subject is the one remaining person in the picture – the dead man on the dissection table around whom the surgeons are gathered.

He is, or was, Adriaen Adriaenszoon, nicknamed 't Kint, 'the kid', twenty-eight years old, and well known to the courts for a string of assaults and thefts over the preceding nine years. In Amsterdam, in that winter of 1631–2, he swiped a man's cape. Unfortunately for Adriaenszoon, his victim resisted, and he was caught. He was tried and sentenced to death by hanging, to be followed by dissection of his body, the usual punishment for serious crimes, dissection having been added specifically to disabuse criminals and their families of any hope they might still cling to of a Christian bodily resurrection. Three days later, on 31 January 1632, his lifeless corpse was taken down from one of the gallows that lined the city's waterfront and moved, ready for the final stage of its punishment, to the city's anatomy theatre.

For in the seventeenth century a dissection was indeed a theatrical occasion. One could only take place when there was a fresh body available, usually from a criminal execution. It would have to happen in the winter months, when the cold would preserve the body long enough for the anatomy to be demonstrated before the stench of decay became overpowering. For many, the chance to witness the criminal getting his just desserts was too good an opportunity to pass up. You might see him hang, and then go along afterwards to the dissection to be sure he was really finished. So among the surgeons and physicians who came looking for instruction and the civic leaders who came to see that justice was done were also those looking for good moral entertainment. Tickets cost perhaps six or seven stuivers (about a third of a guilder, steeper than the price of admission to a players' theatre at the time).

These rare occasions were an assault on the senses. Cold was not enough: incense was burned to disguise the smells coming from the corpse. Music played. Food was eaten, beer and wine were drunk. The magnificent frontispiece illustration in the greatest Renaissance textbook of anatomy, Andreas Vesalius's seven volumes of *De*

Humani Corporis Fabrica (On the Structure of the Human Body) of 1543, shows a dog and a monkey loose among a rowdy-looking crowd. By the time it was over and all of the body parts had been scraped off the table and bagged up for disposal, the gate might have reached 200 guilders or more, enough to pay off the hangman and provide a feast for the members of the surgeons' guild, with a torchlight procession to round off the day.

Rembrandt shows Adriaenszoon's body lying on the table at an angle to us, foreshortened in view. Light pours down on to his barrel chest. I measure him off, and find he stretches more than 120 centimetres from head to toe. Even allowing for the foreshortening, which squashes the criminal up so that he resembles a goblin, he appears to be a powerful man, large and muscular in comparison to the black-coated surgeons. Though partly veiled in a shadow of death, Adriaenszoon's face is open to our gaze. In fact, it looks as if his head must have been propped up on something to allow this surprising indiscretion. His neck, though, which would have been scarred by rope marks from the gallows, remains hidden from view. In contrast to the rosy health of the surgeons, Adriaenszoon's flesh is a pale grey-green. Rembrandt has mixed a tiny bit of lamp-black into his paint to produce this ashen pallor. When Joshua Reynolds saw the painting in 1781, he commented in his travel diary: 'Nothing can be more truly the colour of dead flesh.'

Yet the painting is a fictional construction. In a normal anatomical dissection, the praelector opens the abdomen in order to reveal the major organs, and to allow the most offensive-smelling parts of the digestive system to be shown and then swiftly removed from the scene. Rembrandt gives us Adriaenszoon's trunk intact. Only the left forearm has had the skin pulled away in order to reveal the muscles and tendons beneath. Rembrandt and his client Tulp have chosen to show the criminal's *hand* in dissection. This is a deliberate falsification. Why have they done it?

It is more than likely that Rembrandt was among the audience as 't Kint was gradually disembowelled and dismembered. He may also have found an opportunity to make some quick sketches of Adriaenszoon's body before the anatomy took place. It's possible, too, that he painted in the flayed left forearm and hand from a different subject some time later. Or perhaps he worked from a specimen part long kept in his studio, since an antiquarian who visited the artist shortly before his death in 1669 apparently found in his collection 'four flayed arms and legs anatomized by Vesalius'. There is a suspicion, furthermore, that the *right* hand may not belong to the body we see. Adriaenszoon may have had this hand cut off as a previous punishment for theft, and again Rembrandt may have painted from a different hand. X-ray studies indicate that the right arm of the painted corpse once ended in a stump, and the present manicure is, in the view of some who have examined the painting, 'certainly not that of a thief'.

So not all is as it seems in Rembrandt's early masterpiece. In order for Adriaenszoon's anatomy to take up its position as the subject of the work, his body has had to undergo dual indignities. It has been disassembled by the doctor. But it has also had to be pieced together like Frankenstein's monster by the artist. Both of these actions rely on a new perception of the body, as something that may be opened like a store-room or a treasure chest, an assemblage and a container of mysterious and intriguing parts.

In the years between the publication of Vesalius's treatise on human anatomy in 1543 and Rembrandt's painting of 1632, the topic of human anatomy became something of a craze. The fall of Constantinople in 1453 to the Ottomans saw an influx into Europe of medical scholarship based on Arabic and ancient Greek sources. Restrictions on opening the human body that had prevailed when physicians were also men of the cloth no longer applied. Papal and royal decrees released the bodies of executed criminals for dissection. Suddenly, everything could be 'anatomized', if not physically then at least philosophically. John Donne proclaimed in his *Devotions*: 'I have cut up mine owne *Anatomy*.' The depressive Robert Burton published *The Anatomy of Melancholy*. William Shakespeare had King Lear cry out in his anguish: 'Then let them anatomize Regan; see what breeds about her heart.'

A proper anatomy theatre became essential for any university that hoped to be competitive in the study of medicine. In Protestant areas, these theatres were often converted chapels, indicative not of atheistic medicine taking over from religion, but certainly of the church's acceptance of the new methods of science. This was the case at Leiden University, where an anatomy theatre was built in 1596. Rembrandt grew up in Leiden, twenty miles from Amsterdam, and Tulp studied there, so this was surely the first such space that either of them saw. Today, there is a stylish reconstruction of the theatre at the Boerhaave Museum in the city. It is a circular space with ledges that are steeply raked in order that as many spectators as possible get a clear view of the anatomy that would have taken place on a rotating table in the centre of the room. The Leiden theatre has been adorned with skeletons, both human and animal, including a skeleton man on skeleton horseback, supported on poles from the floor. These macabre embellishments echo a seventeenth-century engraving of the theatre, in which the skeletons mingle with the audience, holding banners proclaiming 'MEMENTO MORI' and 'NOSCE TE IPSUM'. The theatre set up in Amsterdam in 1619 was of similar design. This is where Tulp performs his anatomy. This theatre is long gone today, but the inscription 'THEATRUM ANATOMICUM' remains above the entrance to one of the turrets of St Anthony's Gate, where it was located.

Doctor Tulp, then, was both a pioneer and a leading member of a respectable profession. He had worked to get there. It was as Nicolaes, or Claes, Pieterszoon that he enrolled at Leiden University, the son of a Calvinist linen merchant. He wrote his dissertation on cholera, duly qualified as a doctor of medicine, and returned to his native city of Amsterdam to set up his practice. He was not a specialist in anatomy but a generalist, able to prescribe either medical treatment or surgery for his patients. He adopted the tulip as his emblem, displaying it on his house and coat of arms. The flower, newly arrived from Turkey, would soon become a national mania in the Dutch Republic, sparking the world's first economic bubble as people bid up the price for the latest exotic bulb. But that was a few years away. The doctor was ahead of the game and, as he prospered, his symbol became his name: Doctor Tulip. By 1628, he had risen to the post of praelector in the surgeons' guild. He performed his first public anatomy in January 1631. In Rembrandt's painting, we see him a year later, approaching the age of forty, at the height of his powers, a city alderman, immersed in his second anatomical demonstration.

Tulp's character is the key to the painting's broader message. It turns out to be, as we hope from Rembrandt, rather more than another group portrait. Look at the assembled surgeons' faces, flushed with the January cold and their own importance. Like a cartoon flip book, they move through time from left to right, revealing different stages of expression, from simple perception to intellectual comprehension and finally to something like divine revelation. Tulp himself has the inner light of religious conviction. For, with Rembrandt as his interpreter, Tulp is revealing metaphysical as well as scientific truth. The two men's choice of the hand as the focus of the anatomy demonstrates their true intent. He can show all his dexterity and skill and invention and sleight – as a surgeon, as a painter, as a light-fingers – and then he dies. Man is both vital and mortal; he creates, but he is God's creation.

As if the message is not clear enough, the surgeon standing at the back of the group is pointing at the corpse and looking out of the painting directly at us. He is almost accusing.

It is we who should prepare to receive the lesson.

My own first sight of a body under dissection comes as a surprise. It is a woman. Anatomy texts place unfair emphasis on the male body, not only because anatomists and surgeons were men, but also because male cadavers were what tended to issue from the gallows. The emerging medical profession may have struggled with the constant shortage of bodies for study, but fit young men were not underrepresented among them. Illustrated anatomical texts are filled with well-muscled male youth.

I find other surprises too. The woman I am looking at was obviously of a considerable age when she died. Her skin is the colour of putty, like chicken that has been kept in the freezer too long. The greatest shock is that her head has been cut off, and not where you expect, at the neck, but awkwardly through the chin, because the teeth have been taken away with the rest of the head for dental studies, leaving just the lower part of the skull and jawbone.

She is lying in an unzipped white body bag on one of a dozen steel tables in a dissection room at the University of Oxford's Medical Sciences Teaching Centre. It is a white room, bright with low sunlight that streams in through strip windows and fluorescent lights overhead. Only one thing upsets the air of clinical modernity, and that's the skeletons dangling from frames placed between the tables. Later, in a book, I see a photograph of a university dissection room in Victorian times, and it has the same array of tables and skeletons. The tables are wooden and the bodies wrapped in cloth bandages, but otherwise the scene is unchanged.

I am here as the guest of the Ruskin School of Drawing and Fine Art, accompanying a class of first-year art students. The Ruskin is the only art school in the country that requires its students to draw from anatomy, something that was once a standard part of any artist's training. So I find I am standing more in Rembrandt's shoes than in Doctor Tulp's.

Drawing teaches proper observation. I, too, will attempt to draw what I see. Our instructor, Sarah Simblet, is an artist and an academic. For her doctoral degree she examined the relation between drawing and dissection, the similarities and differences in action of pen and scalpel. She arrives flushed from bicycling in to town. It is another cold January day and her cheeks are as red as those of Tulp and his surgical companions.

Sarah explains that the bodies we will be drawing have been bequeathed by local people for medical research. (The bodies of members of the medical faculty who become donors are carefully sent elsewhere so that they don't end up giving former colleagues a nasty surprise.) There is a disadvantage to this. The sex bias of earlier times is eliminated, but the balance is now tipped in favour of the aged bodies of those who have died a natural death. Those who die young usually undergo post-mortem examination, which means that the body afterwards may be no longer 'viable'.

We pull on white lab coats and latex gloves. Sarah assures us that everything we will see is 'absolutely inert' and there is no obligation to handle or touch the anatomies. The students' anticipation rises. One or two jokily greet each other as 'doctor'. Somebody wonders aloud whether she was sensible to have eaten lunch.

By way of gentle preparation, Sarah shows us first a box of bones. The students have already acquired the rudiments of human anatomy from studying plastic skeletons, but for most of them this is their first encounter with real body parts. 'Help yourselves,' she says, as she picks out her own assortment. She holds up a shoulder blade so thin you can see the light through it, and points out the ridges on the larger bones where the muscles were once attached. I have handled bones before, but still marvel at how light they are.

We move on to the anatomies. The row of tables nearest the window have whole bodies on them in various stages of dissection, prepared by the surgical students of the medical school who work on them progressively over the course of the academic year. The remaining tables support an assortment of torsos and limbs with the skin and layers of subcutaneous tissue cut away in places. These are so-called prosections, dissections of key features of the human anatomy that have been expertly prepared for the instruction of general medical students, who will not perform their own dissections.

We gather round the first table, the one with the old woman on it. Her skin has been cut in such a way that it can be pulled away from the chest to reveal a thin layer of yellow surface fat. We see the muscles connecting the breasts to the ribs. The muscles taper away and become tendons. The tendons, Sarah points out, have 'this rather beautiful silver quality'. The ribs and breast bone have been neatly sawn free from the rest of the skeleton. With a teacher's enthusiasm, Sarah brushes away her long blonde hair from her face, aware perhaps that she should perhaps have tied it back, and peers in to the chest cavity. 'Oh, you're in luck this year,' she says. 'It's beautifully done.' She lifts out the woman's lungs, the right lung with three lobes, the left with two. The spongy tissue has a bluish colour. It suggests that she lived in the country. (City lungs are black, as I find on a later visit to the museum of a London teaching hospital.) Holding them in her hands, Sarah shows how the two lungs come together like the parts of a cast, leaving a void shaped for the heart. Next, she pulls back the pericardium, the cup-shaped membrane that holds the heart in position. We see the heart itself.

The second body is also a woman's, stouter than the first. A thin, rancid odour produced by the creeping oxidation of her body fat pierces the fumes of the fluids used to preserve the anatomies. (Very fat people tend not to be selected for dissection because the fat is simply waste in anatomical terms, and it makes unnecessary work to clear it away.) Her lungs are pushed up high into her chest, perhaps indicative of some medical condition, such as an enlarged liver, or maybe just a normal variation.

The third body is a man's. A tattoo is discernible on his right arm, a motif combining a heart and a sword. He retains his chest hair. This evidence of personal identity makes it harder to see him as just a cadaver. His flesh appears darker than the women's because the blood did not drain easily from his body. He has a large frame, and we observe that his heart has become enlarged so that it could continue to push blood around his clogged arteries.

Seeing these bodies, I am struck by the way the major organs take shape so as to nest tidily alongside one another. This very neatness seems to imply design, and was seen by early anatomists, certainly including Doctor Tulp, as evidence of God's creation. One of my initial naive questions – do organs exist, or are they cultural inventions? – seems to be answered. The organs appear quite distinct from each other and from other tissue. They have their own colours, textures and densities. I can lift them out and pop them back in sequence just as in one of those medical students' plastic models. Sliding a liver in under the diaphragm or tucking one of the lungs round the back of the heart has a pleasing action about it, as the organ and the cavity from which it came feel their way wetly towards one another, slipping easily back into their arrangement in life. After seeing several more bodies, it is clear that people vary on the inside at least as much as they do outwardly. We are not all the same beneath the skin. I see great variations in internal morphology that, seen on the surface, would undoubtedly be cause for comment, disapproval, revulsion and discrimination. Inside the body, they pass unnoticed even by their possessor. What, I wonder, does this tell us about our humanity?

The prosections are a miscellany: a couple of hearts, a ribcage neatly opened up like sideboard doors, a green-tinged gall bladder, kidneys with their ureters linking them to the bladder, a uterus with its ovaries and Fallopian tubes. The intestine is not like the string of sausages shown in cartoons, but is fed by an intricate web of blood vessels along its length. One heart has a plastic tube in it, a relic of historical surgery. One skull has been forced apart along the suture lines where the cranial bones first fuse in infancy. This is done by filling the skull with dried peas and then soaking them so that they absorb water and then gently expand to ease the pieces apart. Some of the prosections have been here for many years. They are preserved in a cocktail of alcohol, formaldehyde and water that pervades the atmosphere with a tangy odour and which still clings to me and my clothes hours after the class has finished. The muscle tissue is more wrinkled than on the dissection bodies. The flesh, I am alarmed to find myself noting, is undeniably the colour of slow-cooked meat. One prepared spinal cord is 150 years old, dissected to a standard that nobody can equal today.

Sarah's course comprises eight classes that take the body piece by piece: shoulder and arm, forearm and hand, torso and so on. I return the second week, which is devoted to the head and neck. From a tank containing a dozen or more, we each choose a head to draw. One old man has a pointed chin with fine white stubble and a Roman nose bent over to one side. His tongue sticks out slightly. He seems full of character, like a gargoyle. The brain has been removed from another head, leaving one eyeball tethered in space, the flesh around the eye socket dissected away. Another man's head has been sectioned vertically, slicing in half even his ginger moustache. He seems quite recognizable from just half of his face, leading me to ponder the importance of symmetry in the human face and form.

The head is a great subject for drawing. There is the formal difficulty of its shapes. There is the hint of identity in the face, or what is left of it. There is the challenge of recovering a sense of life in this now dead flesh. Perhaps it's more than a challenge. Perhaps it is the artist's duty to bring his subject back from the dead in some way. There is too much detail to draw it all. A big part of the artist's job is to edit. What to show? What to leave out? It raises the question of motive. Is the artist depicting an inanimate curiosity, or presenting an allegory of human vanity, or aiming for a scientifically accurate illustration?

The nervous joking stops and the students fall silent as they begin to draw. I have chosen a head and shoulders that lie flat on the table. The face is intact but turned slightly away from me. The skin, fat and some of the muscle of the cheek have been removed, and the neck has been prepared in such a way as to reveal a mass of blood vessels and tendons. I try drawing what I think I see, giving the head an outline and then edging in major features such as the dramatic line where the skin has been abruptly cut away. Aiming for accurate representation, I expect the parts to clarify themselves as I go along, but it doesn't happen. This head, with its tangle of sinews, muscles and tubes, seems hopelessly disordered. I find my freehand pencil strokes can follow the organic curves easily enough. The telephone-shaped opening of a nostril comes out well. But the complex ebb and flow of surfaces and the change of textures between skin and flesh and bone defeat me.

The students draw better, but above all more quickly, than I do. I try to work faster, more spontaneously. I try softer pencils. My efforts disappoint me, as I knew they would. I look at the students' work. A few are plainly not interested, whether in drawing as a medium, or in human anatomy as a subject, but most have responded bravely. I am impressed by their battery of techniques and choice of difficult or apparently unpromising subjects. One, drawing the interior of a skull, cross-hatches dramatically to emphasize the changing fall of light across its concavities. Another seems to find all the loops and whorls in a prosection that includes the aorta, lines I'd never think to follow, and keeps them going to build up an almost abstract composition.

Body parts are complex throughout. They may vary in relative flatness or smoothness, but they don't really have easy bits and difficult bits. In my case, it's all difficult. There is no helpful contrast between organic form and regular geometry as there is in a still life of a bowl of fruit or a pipe on a table.

'Well, I recognize her,' Sarah says diplomatically of my effort when I have given up. She has drawn these same heads herself as well as overseeing her students, and knows them like old friends. In her own work, Sarah has moved on from anatomy to botanical subjects. Her pen-and-ink drawings of trees seem to accentuate their bony aspect, with branches trailing out into sinister fingery twigs and knuckly boles for trunks. She has found the change of subject matter instructive. 'I'm often astonished at the number of errors there are in drawings of bodies in quite well-known texts,' she says. 'You don't get those errors in drawings of plants. With plants there isn't that presumption that we know what this is.'

But the body is us, and we think we know it better than we do. Even Vesalius famously got certain anatomical details wrong, and there is controversy too over whether Rembrandt's dissected hand of the criminal 't Kint is an accurate representation. Leonardo da Vinci, Sarah tells me, drew the valves of the heart correctly in a way that would not be repeated until the twentieth century, but even he couldn't resist putting a hole through the pericardium in order to allow for the passage of spirit, the fluid life force then thought to run through the body.

I am aware as I write about these classes that they may sound more alarming than I found them at the time. It is perhaps shocking to read about wheeled plastic bins labelled so casually with, for example, 'Hands and arms', and to find that they do indeed contain hands and arms, each dissected so as to demonstrate particular points of anatomy, but all then thrown in apparently at random so that they appear to writhe around one another, with a few reaching out of the water in which they are preserved. It could be a gruesome sight. But the context is important – not just the bright light, the clinical surfaces and the antiseptic smell, but also the quiet mood that automatically descends on our group in the presence of the bodies.

I find myself wondering about these people. If they gave their bodies 'to science', were they aware that they might also become artistic subjects? Would they have minded? The Ruskin artists are learning about human anatomy in their chosen way. They are not medical students, but then the body should not be the domain of medicine alone. The way they ultimately employ their knowledge may be different, and less directly applied for human benefit, but they are continuing a noble tradition of forcing the rest of us to see what the human body is like.

If we think of an autopsy, our mind immediately leaps to scenes from television crime thrillers. Here is the detective desperate to get the killer – haggard, blustering, under pressure. And here is the pathologist probing the dead body that will reveal the vital clue – methodical, unflappable, quietly humorous. It's always the pathologist who has the critical distance necessary to solve the case. But it is only latterly that autopsy has come to apply to the inspection of a dead body specifically and almost exclusively for the purposes of discovering the cause of death or disease. The word literally means 'to see with your own eyes'. Its connection with the opened body arises from the first tentative Western investigations of human anatomy in Ancient Greece and the novelty of seeing and knowing the organs and parts for the first time.

Seeing with our own eyes necessarily applies to the body of another. 'NOSCE TE IPSUM' reads the pious placard in the anatomy theatre – 'know thyself'. But we cannot know our own self in this way because we cannot see our own interior exposed. This impossibility allows us to believe in our own immortality. We cannot see ourselves as we are, either inside (because we must be dead first) or out (because we cannot step outside our body to look). So the best we can do is to look at other bodies on the assumption that they are like our own. It's a major step to do this. It requires us not only to accept our own mortality, but also to acknowledge the unity of humankind.

With the privileged specialist at our shoulder to guide us and point out the noteworthy landmarks, we are all able to perform autopsy. As we shall find, the medic and philosopher, artist and writer all have truths to reveal about the human body and its parts.

But first things first. To find our way around, we are going to need a map.

PART ONE

The Whole

Mapping the Territory

On holiday in Greece one time, I remember the ferryman pointing out to me a geographical feature known as Kimomeni Mountain on the mainland across the channel from the picturesque island of Poros. Kimomeni means the Sleeping Woman, and, once you are aware of it, it is impossible not to see her shape in the hills, especially in the evening, when the setting sun sharpens the outline and the retsina kicks in. Her head has clear facial features, her breasts thrust skyward, and her belly tapers away below the lowest rib of her ribcage. Her legs are pulled up so that a knee forms another summit. It's a tourist trap, to be sure, but then again these hills have always looked like this. The Sleeping Woman predates the Acropolis. The ancient Greeks would have noticed her and pointed her out to travellers just as the modern ones do. Even Plato of Athens, thirty miles away, may have remarked upon her.

There are other sleeping women, in Thailand, Mexico and elsewhere, more or less convincing. The Scottish hills known as paps, as well as many others, are likened to human breasts. Lone rocks are named for isolated human figures in myth, such as Lot's wife. In the mountains of California lies Homer's Brow. Open a map of any generously contoured part of the world and sooner or later you are sure to find a feature named after some part of the human form. Such anatomico-geographical similes live on in the age of accurate mapping and aerial views: the 'hand of Michigan' is the mitten-shaped part of that state that juts north, dividing lakes Michigan and Huron.

When looking for an ideal of the human form, though, the Greeks looked not to the land but to the skies. They made man the replica of the universe. In Platonic metaphysics, the macrocosm, which we can translate as 'the large world of order', was answered by the microcosm, the small world of order that was the human body. Parts of the body were assigned to different signs of the zodiac around this time (roughly speaking, moving down the body through the astrological year, from Aries representing the head to Pisces the feet).

The idea of the human body as a microcosm is pervasive and persistent. It occurs in Hindu and Buddhist traditions as well as surviving the transition to Christian belief in the West. The rise of science after the Renaissance, when the anatomists got to work dismantling the mystery of the body, may have dealt a blow to this abstract metaphysics. But it nevertheless continued to hold an attraction for philosophers such as Spinoza and Leibniz. There are echoes of it today in 'new age' thinking and in Gaia theory, which likens the earth to a living organism.

It is no wonder we seek out the body in geography and cosmography. In a real sense, our individual human body is the environment, very nearly the terrain, within which we exist and act out our life. The body is both us and in a way our ecosystem. 'I both have and am a body,' as the sociologist Bryan Turner puts it. Or, as the Stoic philosopher Epictetus pithily told his students: 'You are a little soul carrying around a corpse.' It is this dual mode of existence that makes geography such a compelling metaphor for the body, and that, in turn, gives the body its own potency as a metaphor.

The idea of the body as territory will recur throughout this book. It will be particularly clear in stories of anatomists as they explore the body lying before them like an uncharted ocean, claiming new lands, and naming them after themselves. The Fallopian tubes (the human oviducts) and the Eustachian tube in the ear were claimed and named – by the Italian physicians Gabriele Falloppio and Bartolomeo Eustachi – in the same century as the Magellan Strait and Drake Passage. It also underlies the methodology of what is called topical medicine (*topos* being the Greek for place), which proceeds by isolating problems or diseases within particular parts of the body, and it informs our simple hope that the doctor will be able to 'put his finger on' the trouble, like a place on a map. The display of anatomical specimens in the collection of the Royal College of Surgeons in London has recently been reorganized 'regionally', according to its curator Carina Phillips, rather than by functional connectedness, reflecting a similar change of emphasis in medical education. Books of illustrated human anatomy are still sometimes called atlases, scant change really from the seventeenth century, when they were termed microcosmographias, in evocation of the diagrams of the solar system and constellations known as cosmographias. Both terms hold on to the old idea of the body as the microcosm of the universe.

Why does the geographical metaphor work so well? The body clearly has paths through it, nerves, veins and arteries. These feed certain organs or run from them. Conveying precious fluids, they are like the life-giving rivers worshipped by the Greeks. Simply tracing their routes through the body immediately puts us in mind of a map with distinctive features here and there, and in-between regions where not much happens. After Descartes's philosophical demonstration of the separateness of body and soul, and advances in modern science such as William Harvey's discovery of the circulation of the blood, we could begin to see the body as a kind of machine. But, up until then, it had been a whole world, with parts known and parts yet to be explored, a land with a familiar shore but an uncharted interior.

The human body was also an inspiring prototype. Whole cities as well as individual buildings have been modelled on it. In the fifteenth century, the architect Antonio di Pietro Averlino, known as Filarete, designed the imaginary city of Sforzinda in honour of his patron, Francesco Sforza, the Duke of Milan. It was the first of many ideal city plans of the Renaissance. Sforzinda's walls described an octagonal star for defensive reasons, but within this protective skin the city was conceived for a community that would function as smoothly as the human organism. Zamość in south-east Poland was actually built according to these Italian Renaissance principles: the centre of the city is the stomach – the Great Market; St Catherine's church lies off to one side like the heart; the Zamoyski Palace is the head. There is even the Water Market situated in roughly the same place as the kidneys.

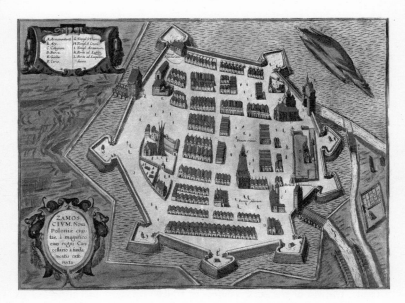

In the following century, the architect and famous biographer of artists Giorgio Vasari laid down his conceptual plan for the ideal palace also in terms of the human frame. The façade was the analogue of the face, the courtyard was the body, the stairways the limbs, and so on. When built, though, these structures did not easily reveal their inspiration. After all, most buildings have a façade to greet visitors and a rear where waste is disposed of without the need for highfalutin anthropomorphic theories. The bodily ideal could be explored more fully in literature. In *The Faerie Queene* by Edmund Spenser, the knights Arthur and Guyon come upon a luxurious castle laid out like a human body. The ascent to the higher storeys is made via the ribs, 'ten steps of Alablaster [sic] wrought'. Reaching the head, the knights find that the mouth is a well-staffed gateway: 'within the Barbican a Porter sate' – that's the tongue – and on either side of him 'twise sixteen warders sat' – the teeth. The eyes are 'two goodly Beacons', while three separate rooms house the various functions of the brain. The first buzzes with flies, representing men's fantasies and imagination. The second contains the intellect and capacity for judgement, while in the last there waits 'an old, old man . . . of infinite remembraunce'.

The poet-turned-priest John Donne, who spent much time considering the body both lustfully and spiritually, imagined it as neither palace nor castle. Instead, he looked into the basements and servant spaces, describing in one of his sermons its 'larders and cellars, and vaults' stuffed with 'pottles [half-gallon containers] and gallons' of urine, blood and other liquids, the body's fuels and wastes.

Gustave Flaubert's friend Maxime du Camp anatomized the city of Paris as a system of organs and their functions, while the socialist philosopher Henri de Saint-Simon dreamed of placing a vast temple in the shape of a woman at the centre of his remodelled utopian Paris. This monumental female saviour of his movement was to have carried a torch in one hand, lighting her kindly face, while the other would have supported a globe containing an entire theatre. Her robes would fall away to a great parade-ground where people could disport themselves in gentle diversions amid the scent of orange blossom. The basic idea is not new at all. As the mythographer Marina Warner notes, a Stone Age temple uncovered at Skara Brae on Orkney adopts the 'cinquefoil form of a schematic female body, the entrance lies through the birth passage'.

Doubtless, there is a Freudian desire to return to the womb lurking in this drive to build habitation patterned after the human body. But, more importantly, at the conscious level of the intellect, the human body was taken as a model for design because it was felt to hold within it an ideal. If man was made in God's image, then should not everything else be made in man's?

The ideal human clearly existed in artists' minds, but could it be described in such a way that all could know it? Could it, for example, be distilled in the new language developed by the Greeks – mathematics? Plato believed that sight was the noblest of the five senses. His notion of human beauty was therefore a visual one and has set the terms for the philosophical discussion of beauty ever since. Indeed, it continues to govern our prejudices today, as we see when a television presenter is suddenly sacked for no greater sin than simply growing old at the same rate as the rest of us. Fair or not, though, the measure of beauty we are looking for must also be visual.

In the latter half of the fifth century BCE, the Greek sculptor of athletes, Polykleitos, set out a prescription for human beauty in a text that he called the *Canon*, and used these proportions to create an exemplary bronze nude of a young man carrying a spear, the *Doryphoros*. The original sculpture does not survive, but a few fragments and Roman copies in marble have been enough for faithful casts to be made, which are widely scattered in the world's museums. The artist concentrated on the torso, giving it magnificent pectorals and obliques – the muscles just above the hips. The second of these features appears especially overdeveloped to modern eyes, but it helped to define the ideal in Classical sculpture, becoming known as the girdle of Achilles or the Greek fold. This spear-carrier is perhaps the most copied statue in antiquity. His chest was even taken as a model for the body-fitting bronze armour of generations of later Greek and Roman soldiers. This 'muscled cuirass' (or *cuirasse esthétique* in French) replicated not only warrior-like features such as the pectorals and the ribs, but also the navel and even the nipples. This ideal survives in comic-book heroes such as Superman and Batman, whose tight tunics reveal every muscle – but generally omit these homoerotic extras.

Like the original *Doryphoros*, the text of the *Canon* is lost, and with it presumably the numbers that would reveal to us Polykleitos's system of ideal proportions. Surprisingly, perhaps, it has proven impossible to deduce them from the sculpture itself. The human body has simply too many dimensions, and too many points where one might put one's measuring tape, to begin to guess.

The system of human proportion laid down some 400 years after Polykleitos by the Roman architect Vitruvius has fared better. It is found in the only surviving work on architecture from the Classical period, *De Architectura*. Vitruvius's ten volumes were the standard text for architects until the Renaissance, when men such as Leon Battista Alberti and Andrea Palladio brought out their own multi-volume guides. Vitruvius's ideal template for the human body comes in the third volume of the set, which covers his principles for the design of temples:

> no building can be said to be well designed which wants symmetry and proportion. In truth they are as necessary to the beauty of a building as to that of a well formed human figure, which nature has so fashioned, that in the face, from the chin to the top of the forehead, or to

the roots of the hair, is a tenth part of the height of the whole body. From the chin to the crown of the head is an eighth part of the whole height, and from the nape of the neck to the crown of the head the same. From the upper part of the breast to the roots of the hair a sixth; to the crown of the head a fourth. A third part of the height of the face is equal to that from the chin to under side of the nostrils, and thence to the middle of the eyebrows the same; from the last to the roots of the hair, where the forehead ends, the remaining third part. The length of the foot is a sixth part of the height of the body. The fore-arm a fourth part. The width of the breast a fourth part. Similarly have other members their due proportions, by attention to which the ancient Painters and Sculptors obtained so much reputation.

Vitruvius's scheme enabled all the major features of the human form to be described in terms of simple ratios of just a few small numbers. Four digits made a palm and six palms a cubit. The height of a man was four cubits or six feet. By means of some elaborate argument, Vitruvius even managed to suggest that 'numbers had their origin from the human body'. Today, we might sceptically observe the tricks by which Vitruvius has contrived his elegant system, picking human features to suit his ratios rather than those anybody might use – the underside of the nostrils rather than the tip of the nose, the eyebrows rather than the eyes, and so on.

Vitruvius goes on to explain that the navel is 'naturally placed at the centre of the human body', and that a circle drawn about that centre on a man with his arms and legs outstretched will touch both his fingers and his toes. Similarly, he writes, the arms at their full horizontal extent span four cubits, the same as the height, so that a square can also be drawn around the body. Both of these shapes – the circle and the square – were symbolically important in the design of temples because of their geometric purity, and it was important to connect them with the human figure in order to demonstrate its divine proportion.

Vitruvius left no illustration to accompany his detailed text. While several artists contributed illustrations to editions of *De Architectura* published in the sixteenth century, they had trouble reconciling all of Vitruvius's required elements – the man's dimensions, the square and the circle. Assuming that the square and the circle must be concentric, they were forced to distort the human figure to fit them both. It was only Leonardo da Vinci who was able to reconcile all of the Roman architect's precepts in a single harmonious design.

Leonardo was probably the first artist to cut up the human body and draw what he saw. He boasted of dissecting more than ten bodies, proceeding in stages as each body gradually decayed, and repeating the whole exercise in order to gain an appreciation of the typical differences between one body and another. He described the experience in his notebooks, thrilling his readers with his account of spending 'the night hours in the company of these corpses, quartered and flayed and horrible to behold'.

The genius of Leonardo's solution was to take the truth of the human form as his ultimate guide and make the geometry fit around that. So he simply superimposed the standing man in the square on top of the man with limbs outspread in the circle, a delightful resolution in which the twin sets of limbs even bring a suggestion of human animation. The result is that both the square and the circle touch the ground. The figure inscribed in the circle has his navel at the centre, as Vitruvius demands, but as the centre of the square now lies lower than the centre of the circle, this coincides not with the navel but, significantly, with the genitals. Here, then, is the human figure: progenitor and progeny, creator and creation. It's undeniably neat. But is it true? After all, why should the body be describable in terms of simple numerical ratios?

Late in his career, the twentieth-century Swiss architect Le Corbusier felt the need to reinvent Vitruvian man for the modern age. If he hadn't chosen architecture, Le Corbusier might have been a boxer. He drew sketches of boxers and likened himself to a boxer in his professional struggles. His new ideal man, which he called Le Modulor, raises a massive fist to the sky. The first version of Le Modulor was based on a typical Frenchman of 1.75 metres, but the architect disliked the metric system, based as it is on the dimensions of the earth rather than on the human body, and later announced that his model would henceforth be a six-foot Englishman, 'because in English detective novels, the good-looking men, such as policemen, are always six feet tall!' To the top of that raised fist, Le Modulor is 226 centimetres high, and his navel is centrally positioned at 113 centimetres. The distances from the ground to the navel and from the navel to the top of the head are in the proportion of the golden ratio (0.618:1 = 1:1.618), as is the remaining distance from the head to the fist, and he is indeed six feet (182.8 centimetres) tall. This personal system of proportion devised in the 1940s governs the proportions of the Unité d'Habitation, Le Corbusier's influential apartment block in Marseille. The Modulor logic yields a building that is 'in every way as keen, sharp and terrifying as the Parthenon', in the words of one architecture critic, even if it is a 'semantic strength' that is gained rather than anything magical to do with the numerical ratios employed or their intrinsic connection with the human body. Le Modulor man features as a sculptural relief in the concrete, but it seems unlikely that many of the residents know their comings and goings are overseen by an English policeman.

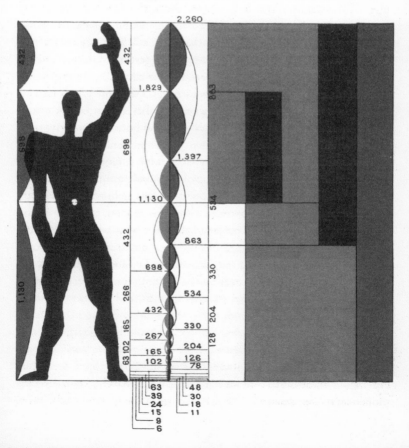

Le Modulor may seem a bit arbitrary, perhaps even gently mocking the notion of the ideal body, but like Vitruvian man he is sure of his height. For it is certainly true that, in order to approach any kind of human ideal, a person must not only have harmonious proportions, he or she must also be the right size.

Ancient units of measure were directly based on the dimensions of the human body. We still use quite a few of them today. One defin-ition of the inch was the length of the (royal) thumb from tip to first joint. The foot covers a narrow spread of distances based on the lengths of human feet, among which our imperial twelve-inch foot is about average. As well as being the width of four palms, a cubit is the length from the elbow to the longest finger tip, and is usually equated to eighteen inches, but sometimes to twenty-one or more. An ell, derived from the Latin for elbow, *ulna*, was originally the same as a cubit. Later, the ell was adopted as a unit for measuring cloth at a much longer forty-five inches. It may have taken the name because measuring out involved holding the cloth in the fingers of one hand up to the opposite shoulder and then straightening the elbow until the arm is outstretched on its rightful side. At any rate, I am a little shorter than Le Corbusier's English hero, and this distance comes to nearly forty-five inches on me.

All of these units of measurement start from a different part of the body. There is nothing that says they have to be in simple arithmetic relationship to one another. So the fact that they have been put together in just this way – twelve inches making exactly one foot, for example – is evidence that Vitruvius and his mathematizing idealism has stayed with us.

These measures, however, are all linear. The body is less helpful when it comes to areas, masses and volumes. A few units are based not on man's dimensions but on his capabilities. An acre is the area that a man and his ox were supposed to be able to plough in a day, for example. But quantities such as the weight that a man can lift or the volume of water (or beer) he can drink are so variable that the body no longer provides a good standard. Even Vitruvian six-footers can be scrawny or stout. Nevertheless, there is a smattering of other so-called anthropic measures, even including units of time. In Hindu tradition, *nimesha* is the length of a blink of an eye and *paramaanus* is the interval between blinking.

The importance of being the right size is made apparent in stories where it all goes wrong. *Alice in Wonderland* and *Gulliver's Travels* are two of the best-loved tales where the relative size of the central character becomes important. Alice and Gulliver undergo instructively different experiences. When Alice first drinks from the bottle labelled 'DRINK ME' and finds herself shrinking, and then eats the cake that makes her grow again, she extrapolates from what is happening to her to imagine the horrible consequences of being taken to the extreme of either scale – perhaps 'going out altogether, like a candle'. Alice is in a world where measurement – and even number itself – can no longer be relied upon, as she finds when she tries to recite her multiplication tables and they too come out wrong.

Gulliver's measure of things, on the other hand, remains rock-steady. He confidently describes the tallest trees in Lilliput as 'seven foot high', while in Brobdingnag one of the reapers who find him hoists him aloft 'above sixty foot from the ground'. The reader understands from this that Gulliver retains his proper size. Calculations are a feature of *Gulliver's Travels*, too, most notably when the Lilliputians calculate that they must feed Gulliver 1,728 times as much food as they require for themselves, since Gulliver is twelve times their size in all three dimensions (and twelve cubed is 1,728).

Alice clearly changes size down the rabbit hole, whereas Gulliver visits different-sized lands. In both stories, though, the rule of law is strenuously asserted. 'You've no right to grow *here*,' the Dormouse admonishes Alice as she begins to resume her proper scale in preparation for her return above ground. 'Rule Forty-two. *All persons more than a mile high to leave the court*,' barks the King. The Emperor of Lilliput likewise imposes conditions on Gulliver: 'First, the Man-Mountain shall not depart from our dominions, without our licence under our great seal.' Size matters, and being the wrong size calls for disciplinary measures to bring the offender back into line.

Today, we have largely relinquished the notion of ideal man. The eighteenth-century painter and cartoonist William Hogarth declared that it was impossible to find geometry in human faces, and celebrated their irregularity in his satirical caricatures. In a section headed 'Proportion not the Cause of Beauty in the Human Species' in his famous essay of 1757, *A Philosophical Enquiry into the Origin of Our Ideas of the Sublime and Beautiful*, Edmund Burke refuted the whole concept, pointing out that 'ideal' proportions could be found in people judged beautiful and ugly alike. 'You may assign any proportion you please to every part of the human body; and I undertake that a painter shall religiously observe them all, and notwithstanding produce, if he pleases, a very ugly figure.' He reserved special criticism for Vitruvian man. Man was never based on a square; he was, if anything, more like its opposite, a cross. 'The human figure never supplied the architect with any of his ideas.'

Thanks to a Belgian pioneer of statistics and social science, Adolphe Quetelet, we have now instead the idea of 'average' man and woman, a development in thinking about humanity that required the invention of statistics, with its concept of mean (average) and standard deviation (the extent of the variance either side of the average). Quetelet was the first to gather systematic data on human height and weight, introducing the concept of the average man (*l'homme moyen*) in a book of 1835. Quetelet even found a way to decouple the two measures, so that people could be usefully described as heavy or light *for their size*, introducing the index now named after him, better known to most of us as the body-mass index.

Quetelet's new approach gave licence for a vast exercise in data collection. The field of study that emerged was christened anthropometry a few years later. In recognizing that one person was physically different from another and that measuring those variations might yield useful information, the anthropometrists implicitly acknowledged that one human was as valid as another and so in effect rejected the concept of ideal man.

Such data was too powerful to be left solely in the hands of scientists. In the Museum of the Prefecture of Police in Paris is a reconstruction of an unusual photographic studio. In addition to the huge cameras of the day, it contains an assortment of calipers and rulers and other paraphernalia for measuring subjects as well as capturing their image. This was where Alphonse Bertillon introduced the world's first scientific identity cards. In addition to frontal and profile photographs, Bertillon's cards gave major body dimensions – and some surprising minor ones, including sixteen characteristics describing the shape of the ear. He tried them out on members of his family. His own card, made on 14 May 1891 when he was thirty-eight, shows him with a trim beard, short wiry hair and a high forehead, his head seeming a little too large for his body. In fact, we can read from the card that his head was 19.4 centimetres tall, while his height from the waist was 78 centimetres and his chest was 95.2 centimetres around. His left foot measured 27.4 centimetres. Bertillon, curiously, came from a family that seems to have had a genetic predisposition to this sort of work: his elder brother was the director of statistics for the city of Paris; his father founded its school of anthropology; and his grandfather had developed the work of Quetelet and coined the word demographics. Bertillon's innovations – he also introduced crime-scene photography – saw him rise from a lowly clerical position when he joined the Paris police in 1879 to lead its influential Judicial Identity Service less than a decade later. 'Bertillonage' was soon taken up by police services around the world. Although it could not be used to establish definite guilt, as further persons not known to the police might have similar measurements, Bertillon's method was nevertheless good enough to rule out suspects from police enquiries if they did not match a witness's description.

It was not possible to prove guilt by using body measurements until the discovery that fingerprints are unique to each individual. After the Indian Rebellion of 1857, William Herschel, a British colonial administrator in Bengal, made himself even more unpopular than he doubtless was already by requiring local workers to guarantee their contracts with a handprint. Herschel also recorded his own fingerprints over a period of years, showing that they did not change. His work was noticed by Francis Galton, one of the foremost figures in Victorian science. Even by the standards of the age, Galton was obsessed with measurement. Over the course of an indefatigable career spanning seventy years, he made many contributions to science, including drawing up the first weather maps, questionnaires and intelligence tests. He invented a handheld 'pocket registrator', a bit like the devices used by aircraft flight attendants to count passengers, which could track five independent variables at once, according to which buttons you depressed. The journal *Nature* noted that it would enable scientists 'to take anthropological statistics of any kind among crowds of people without exciting observation'. Galton simply could not rest. One of his papers was titled 'Notes on Ripples in Bathwater'. Another time, in a dull lecture at the Royal Geographical Society, he sought to derive a quantitative index of human boredom from the rate of fidgeting among members of the audience. His true legacy was not in any one thing that he measured, but in the advances he contributed to the methods of statistics needed to process all his data.

Galton studied the prints Herschel had made along with prints from other subjects, using a pantograph he had built for measuring moths' wings to trace and magnify key details. He noticed that no two fingerprints appeared to be the same, but was able to go further than this and confirm their uniqueness by statistical analysis. Galton had corresponded with Bertillon – both men proudly carried their own Bertillon system identity cards – and had been influential in recommending Bertillonage to British police forces. Fingerprints had occasionally been used, like the other measurements made by Bertillon, as a means of disproving a suspect's connection with a crime. But now Galton saw that fingerprinting was actually a far more powerful technique that could be used for catching criminals. In 1902, Rose Guilder, a parlourmaid, noticed a thumbprint in new paintwork following a burglary in the house where she worked. It was the first time that fingerprint evidence was brought to court. Galton, meanwhile, pursued his own research agenda, collecting thousands of prints in a futile hope that he might be able to use them to demonstrate people's relatedness.

Galton held an ardent admiration for his cousin Charles Darwin. (Possibly it is not a coincidence that among his many books is one titled *Hereditary Genius*.) But where Darwin studied the animal kingdom, Galton focused on his fellow man. And woman. Travelling through Africa with a party of missionaries as a young man in 1850, he was startled to observe the wife of one of the party's interpreters, 'a charming person, not only a Hottentot in figure, but in that respect a Venus among the Hottentots'. Naturally, he wished to obtain her measurements. But there was a difficulty. 'I did not know a word of Hottentot, and could never therefore have explained to the lady what the object of my footrule could be.' He dared not ask the interpreter to negotiate for him. Yet there she was, 'turning herself about to all points of the compass, as ladies who wish to be admired usually do'. Then Galton realized that his instruments held the solution to his dilemma. He picked up his sextant and, standing at a respectable distance, recorded 'her figure in every direction, up and down, crossways, diagonally, and so forth, and I registered them carefully upon an outline drawing for fear of any mistake; this being done, I boldly pulled out my measuring tape, and measured the distance from where I was to the place she stood, and having thus obtained both base and angles, I worked out the results by trigonometry and logarithms'.

In 1884, Galton set up a laboratory at the International Health Exhibition held at South Kensington in London, and gathered data from volunteering visitors on their 'Keenness of Sight and of Hearing; Colour Sense, Judgment of Eye; Breathing Power; Reaction Time; Strength of Pull and of Squeeze; Force of Blow; Span of Arms; Height, both standing and sitting; and Weight'. He used the new technique of photography to make 'composite' portraits, layering up individual exposures to produce a supposed average. In this way, he sought – vainly, again – to distill the typical appearance of many diverse populations. All in all, Galton's anthropometric project was far-reaching, and we shall hear more from him in later chapters.

Scientists do not need misleading syntheses like Galton's composites, but they do need typical specimens. Zoologists keep one specimen of every animal, which they call the holotype of the species. It is the benchmark against which other specimens are compared to see if they belong to that species or some other. The scientist who first described the species has the privilege of selecting the holotype. These holotypes are scattered through the university museums of the world.

So where is the human holotype? For that matter, who is the human holotype? Oddly, there isn't really one. This is partly because holotypes are only a designated requirement for species described since 1931, and partly because there is no scientific ambiguity about membership of the human species. (Racists might disagree, but their objections arise in large part because different races can interbreed, which demonstrates our common humanity.) In 1959, however, the Swedish naturalist Carl Linnaeus was nominated for the position, even though he had been dead for 181 years. Linnaeus's *Systema Naturae* of 1758 introduced the nomenclature for species that we still use today, and included his description of our species, with the new name he gave it of *Homo sapiens*. He has not been the only candidate. More recently, a story emerged that the American palaeontologist Edward Drinker Cope put himself forward for the job. Shortly before his death in 1897, Cope sold his fossils to the American Museum of Natural History, and instructed that his own remains were to be preserved in aid of this unusual bid for immortality. The exercise may have been a last hurrah in the scientist's battle with his palaeontological rival in the 'bone wars', Othniel Charles Marsh, as he also wished that his brain be weighed to see if it was more massive than Marsh's – a challenge which Marsh failed to take up. Cope's bid failed because his story only came to light much later, when, unknown to his latterday backers, Linnaeus had already been adopted in the post – although his bones are likely to remain undisturbed in their grave at Uppsala in Sweden.

The constant search for a standard reference image of the human body ends for now with something called the Visible Human Project. We have come a long way from Vitruvius and Polykleitos. And today both man and woman are presented – although, as usual, man came first.

The Visible Human Project began in 1988 as an initiative of the United States National Library of Medicine in response to the rise of two new technological possibilities: first, the ability to freeze human tissue without damaging it; and second, the rise of digital image processing. The idea was to take a human cadaver, slice it up and then photograph it to put together the first detailed visual reference of the human interior based on an actual body.

As with the anatomized bodies painted by Rembrandt and others long before, the chosen subject was a convicted criminal. Joseph Paul Jernigan of Waco, Texas, was executed in 1993 by injection with a lethal dose of potassium chloride, twelve years after being sentenced to death for burglary and murder. Prompted by the prison chaplain, and unable to donate his organs for transplant as they would be poisoned by the potassium chloride, he signed a consent form to donate his whole body. Jernigan passed the 'audition' because he suffered from no disfiguring disease and had not undergone major surgery, either of which would have made him anatomically unrepresentative. The authorities must have been keen to press on with the project, though, because Jernigan was not quite ideal, having had an appendectomy and missing a tooth. Within hours of his execution, Jernigan's body was flown to the University of Colorado and recorded as a set of magnetic resonance images for reference. Then it was frozen and scanned again. Once solidified, the body was sliced sequentially in planes parallel to those used in the MRI scans, one millimetre at a time, and the exposed sections were photographed. The tissue shaved off each time was reduced to 'sawdust'.

The National Library of Medicine put the images on a website in November 1994. The overall view of Jernigan shows a moderately overweight man with a shaven head and a short, thick neck. He is heavily tattooed and highly recognizable. The sections through his body, on the other hand, are baffling to the untrained eye. Each looks like a massive chop in a butcher's shop. It is hard to discern even major organs amid the dark red tissue, in marked contrast with the prepared cadavers I had seen in Oxford. The effect of reducing the three-dimensional complexity of the human body to a series of flat planes is once again to remake the body as a kind of map, with nameless islands of red in a sea of yellow fat.

A female visible human was added a year later. She remains anonymous, known merely as a 'Maryland housewife', who died of a heart attack aged fifty-nine. She has a rather square head with a broad mouth and rounded chin. She too is almost neckless. The National Library of Medicine anticipated that the Visible Human Project would mainly benefit medical students, but uptake has been far

wider, with many others finding the idea of visualization too powerful to resist and going on to produce their own fly-throughs along
blood vessels or atlases of parts relevant to their own specialisms. It
has generated popular interest, too. The media and even scientists
involved with the Visible Human Project often refer to the two subjects as 'Adam' and 'Eve'. 'Adam' has had the most coverage, because
he came first, because his nefarious life is known to us and because of
the belief that he might have earned a kind of redemption by giving
his body indirectly to save other lives. Destroyed as a consequence of
his punishment, he has been digitally reconstituted – almost reincar-
nated, in the literal sense of that word, meaning 'restored in bodily
form'. Such narratives are absent from the visible woman. Her untold
story is this: she is the more scientifically valuable of the pair.
Recorded later, she was cut into thinner slices – three times as many
in all – to yield a more detailed library of images. Almost biblically,
however, it is 'Adam' who continues to be the primary reference.
Most mainstream research has been based on him, while 'Eve' has
been 'primarily used for reproductive anatomy'.

According to Lisa Cartwright, an American expert on visual culture and gender studies, the Visible Human Project 'stands a strong
chance of becoming the international gold standard for human anatomy in coming years'. It is far more than just a visual record. Its sliced
and reassembled human bodies can be experienced and manipulated.
They provide an immersive virtual environment. Naturally, the
dream now extends to 'animating' the bodies.

Nevertheless, the Visible Human Project has its flaws. Because it
presents the internal body as it is, it is paradoxically not always a useful teaching aid. The sheer density of detail makes it hard to pick out
what matters. It is a complement to, not a substitute for, the neat,
colour-coded diagrams of medical textbooks. The way the data is
organized in horizontal slices through the body conflicts with what
trainee doctors will see later in clinical medical images, where the
plane of the image may be at a different angle, or the body positioned
in a different way, and so on. In a strange way, these images may have
more to say to the lay person. They give us a new view of ourselves
as we really are.

In this sense, the Visible Human Project may be seen as the antithesis of the better-known Human Genome Project. Whereas the decoded human genome yields an inscrutable list of letters and numbers describing the thousands of genes and the exact sequence of billions of amino acids that comprise human DNA, the Visible Human Project shows us two real people. According to the Australian social scientist Catherine Waldby, each aspires to be, and is in its way, an 'exhaustive archive of human information', but only the Visible Human Project is 'spectacular'. And, if, as Wittgenstein tells us, 'the human body is still the best picture of the human soul', then it is perhaps the best answer we have yet to the long-held urge to visualize the self.

Now, let's take our own slice of human flesh.

Flesh

How much is a pound of flesh?

To Shylock in *The Merchant of Venice*, it is beyond price: 'The pound of flesh which I demand of him / Is dearly bought, 'tis mine, and I will have it.' The merchant Antonio, you will remember, is strapped for cash while he waits for his ships to come in. He has agreed nevertheless to support his impecunious friend Bassanio in his plan to travel to Belmont, there to woo the lovely (and rich) Portia, and has sent him off to raise the necessary 3,000 ducats, for which he, Antonio, will stand bond. Bassanio finds the Jewish moneylender and they agree terms. Unusually, Shylock asks for no interest, but demands instead the forfeit of a pound of Antonio's flesh if he proves unable to repay the loan. Shylock and Antonio are enemies and business rivals, not least because Antonio undercuts Shylock's usury by lending money to his friends interest-free, as Christian doctrine demands. When the loan comes due three months later, Antonio is indeed unable to repay it, thinking his ships to be wrecked, and the unhappy matter comes to court. In desperation, Bassanio offers Shylock his capital back and the same amount again, a total of 6,000 ducats (the money suddenly available from his betrothed Portia). But Shylock haughtily refuses six times as much. 'I would have my bond,' he insists.

What of this pound of flesh in physical terms? Is its removal supposed to be survivable? Shakespeare has his characters consider the matter in some detail. It is clear in the play that it is Shylock himself who is to wield the knife – Jews were some of the best surgeons and anatomists of the day. But where will he bring it down? When terms are agreed, Shylock stipulates that the flesh is 'to be cut off and taken / In what part of your body pleaseth me'. In court, though, he is told by the 'doctor of law' brought in to adjudicate on the matter (in fact Portia in disguise) that the flesh is 'to be by him cut off / Nearest the merchant's heart', with the contradictory injunction added for him to 'Be merciful'.

The pound of flesh is not Shakespeare's invention. He may have got it from 'Englished' Italian sources or indirectly from the four-teenth-century *Cursor Mundi*, written in Northumbrian dialect. In this version, the Jew, brought to the court of one Queen Ellen, vows to take his victim's flesh in the most hurtful manner possible, by cut-ting out the eyes, hands, tongue, nose 'and so on until the covenant be fulfilled'. The forfeit has echoes of legally sanctioned punishment by amputation.

It is always hard to estimate how much any part of the body weighs since it is for most normal purposes inseparable from the whole. But it is possible to get some sense of what a pound of flesh might amount to. Human and animal flesh are of roughly equal density, so a pound of beefsteak gives a good visual impression. A more memorable method is to dunk your hand in a brimming bucket of water until the displaced liquid weighs this amount (water also being about the same density as the human body). In my case, I find the chop comes a couple of inches above the wrist. Alternatively, a pound would take off most of a man's foot. Of the organs I was able to handle at the Ruskin School, the heart came closest to the required weight. A dissected heart weighs about two-thirds of a pound. Dripping with fresh blood, it might weigh a pound.

Yet Shylock is told he may not take the heart, merely the flesh around it. In general, then, flesh is characterized by what it is not. It is not the organs, which do particular jobs in the body. In animals, flesh is used to mean the edible meat, apart from the offal (the meat allowed to *fall off* the butcher's block). Neither is it hard bone. The biblical phrase 'flesh and bones' implies that flesh is soft. 'Flesh and blood' – a phrase Shakespeare uses many times in his plays – mean-while suggests that flesh is solid in contrast to running blood. Although it may on occasion be synonymous with the skin, flesh is not skin either. Flesh is also distinct from 'spirit'; indeed the two are opposed in constant moral battle. The flesh then is the physical bulk of the body, principally muscle but also the fat. Flesh has depth. We may cut into its thickness. We imagine it in three dimensions. In his celebrated essay 'On the Cannibals', Montaigne writes vividly of tribes who might roast a captured enemy and then send 'chunks of his flesh to absent friends'.

We never find out which chunk of Antonio's anatomy is to go, of course. Quick-thinking Portia examines the letter of the contract, and observes that it specifies a pound of flesh, no more and no less. She rules that Shylock may have his pound of flesh, provided he sheds not 'One drop of Christian blood', and takes an exact pound to within a twentieth of a scruple (a scruple was little more than a gramme).

This judicial pronouncement is meant to pose a moral conundrum, not merely a dissector's dilemma. The lawyer's interpretation follows biblical convention in generally distinguishing flesh from blood. In Jewish doctrine, the flesh is the body (they share the Hebrew word *bâsâr*). But then, as Leviticus tells us, 'the life of the flesh is in the blood'. So there is an important distinction to be made between the two. Where 'flesh and blood' appear yoked together in the Bible it is usually in reference to burnt offerings and animal sacrifices. Because his bodily flesh may be taken but not his vital blood, we understand at least that Antonio is not to be sacrificed in this brute fashion.

Bodies and their parts abound in Shakespeare. 'Flesh' occurs 142 times, with *The Merchant of Venice* employing the word twice as much as any other play. There are 1,047 'heart's in the plays and sonnets, with another 208 'heartily's, 'sweet-heart's and other variations. *King Lear* has the highest count with thirty-nine, not *Romeo and Juliet* as you might expect. 'I cannot heave my heart into my mouth', Cordelia answers viscerally to her father's demand to know whether she loves him any more than her voluble sisters do. There is even a subtle indication that she truly is her father's dearest daughter in her name: Cordelia, Shakespeare scholars have noted, is homonymous with cor-de-Lear (heart of Lear).

By his own admission, Hamlet is 'pigeon-liver'd, and lack[s] gall'. The Dane also accounts for the single occurrence of ankles in Shakespeare, when he appears before Ophelia, 'his stocking fouled, / Ungart'red, and down-gyved to his ankle; / Pale as his shirt, his knees knocking each other'. Macbeth speaks of his 'barefaced power' – the first English usage of the adjective. 'Lily-liver'd' is Shakespeare's coinage too, used twice, in *Macbeth* and *Lear*. A pale liver was thought to be a sign of weakness, related to its then presumed role in generating blood and bodily heat. There are heads and hands, eyes and ears by the hundred, but more significantly also 82 brains, 44 stomachs and 37 bellies, 29 spleens, 20 lungs, 12 guts, 9 nerves and a lone kidney, which crops up in *The Merry Wives of Windsor*, when Falstaff seeks to paint himself as a pitiable figure as he recounts the indignities he has suffered at the hands of the 'merry wives' – 'a man of my kidney', as he splutters incredulously. Indeed, no character in Shakespeare is more splendidly corporeal than Falstaff, who has in this same scene already reminded us how, in the course of one of the women's tricks, his vast, collapsing form was 'carried in a basket, like a barrow of butcher's offal'.

Shakespeare was writing at a time of crisis in the development of our understanding of the human body. It was around this time that the body was given, as it were, a hard outline in contradistinction to the rest of the world. We became *homo clausus*, as the sociologist Norbert Elias labels us: closed-off man. I'm not entirely sure I buy this theory. Surely the living body has always been an impenetrable mystery. When I scratch because I have some itch below the surface, I know its cause will remain hidden to me by my skin. And so it was always. I am tantalized by the thought that if only I could see through it, just briefly part it even, then I could deal with the problem more effectively. Doctors must feel this frustration still more keenly. Yet this is apparently a modern thought. According to the theorists, it simply was not within the imaginative compass of itchy medievals to think in this way. They would have sought their answers to the hidden body's ailments exteriorly, perhaps by looking to astrology and magic.

The rise of anatomy is part of this shift, for the urge to open up the body demands that it is closed to begin with. The anatomist, like the sceptic, must see with his eyes in order to believe and understand. Vesalius's *De Humani Corporis Fabrica* threw open the doors to this inner world. People began to speak more boldly and unashamedly in bodily terms. Even Queen Elizabeth assured her troops preparing to repel the Spanish Armada: 'I know that I have the body but of a weak and feeble woman, but I have the heart and stomach of a king, and a king of England, too.' Shakespeare's abundant references not only to external parts of the body but to the innards that we so rarely see are the writer's response to new literary possibilities. The body's parts provide a wealth of fresh images and metaphors. The Italian historian of medicine Arturo Castiglioni even makes the claim that Shakespeare got the idea for his most famous visual scene, where Hamlet in the graveyard picks up the skull of the king's former jester and holds it in his hand while speaking the lines 'Alas, poor Yorick!', from one of the illustrations in Vesalius, which shows 'a skeleton in meditation', with its right hand resting on a skull placed on the stone tomb in front of it.

Shakespeare goes further than his contemporaries into this new world of language. He is medically literate, and includes somewhere in his plays references to most of the diseases and remedies of the day. More than this, his use of corporeal images encourages our involvement in the drama and produces in us a strong identification with his characters. This distinguishes him from his contemporaries such as Christopher Marlowe, Ben Jonson and even the bloodthirsty John Webster. And, of course, the new language and juicy metaphors based on body parts could only work dramatically if Shakespeare's audiences already shared his sense of the human body.

It is Hamlet who wrestles most with the meaning of human embodiment, using successive scenes to probe the question ever more deeply. Is the embodied self bounded by the physical edges of the body? Upon what he calls Hamlet's 'transformation', his uncle Claudius, the new king, observes that 'nor th' exterior nor the inward man / Resembles that it was'. Hamlet says of himself: 'I could be bounded in a nutshell and count myself a king of infinite space were it not that I have bad dreams.' As it is, he struggles to reconcile the confines of his body with the scale of his increasingly crazy ideas. Hamlet dreams: 'O, that this too too solid flesh would melt.' And in his most famous soliloquy, he weighs the possibility of ending for ever 'The heartache, and the thousand natural shocks / That flesh is heir to'.

In *Macbeth*, it is images of blood that predominate. Blood slops and surges through the play like a river bursting its banks. No longer properly contained within the body, it stains daggers and hands and faces. It even spills out of the drama itself and into the real world of the theatre to 'Threaten this bloody stage', as one character announces. The witches stir baboon's blood and sow's blood into their cauldron. By Act Three, Macbeth is in so deep he finds he must 'wade' in blood. Scotland, like Denmark, is a body: 'Bleed, bleed, poor country!' says Macduff. 'It weeps, it bleeds,' concurs Malcolm a few lines later.

Almost equally liquid imagery accompanies Falstaff – that barrow load of 'butcher's offal' – through the action of three plays. In *Henry IV, Part I*, the fit young Prince Henry repeatedly taunts Falstaff about his alarmingly mobile insides: 'you carried your guts away', 'that stuffed cloak-bag of guts', 'how would thy guts fall about thy knees!' Again, the two characters represent facets of the body politic, presently soft and flabby, but with the potential to become lean and efficient. We hear the same language today, for instance from fiscal conservatives who routinely refer to state budgets as 'bloated'. Indeed, it seems doubtful whether a conspicuously fat person could be elected as a national leader today, even in countries where obesity is epidemic among the electorate.

Before closing the lid on Shakespeare's body, we should pause to consider 'this mortal coil', the most famous vital image of all in the most famous speech of all, Hamlet's 'To be, or not to be' soliloquy. What is it? Shakespeare's strange and powerful phrase naturally suggests many things. The word coil meant turmoil or trouble in the sixteenth century. A coil was a colloquial term for a noise and bustle, derived from its original meaning as a verb to heap up, gather or collect, from the French *coillir*. Yet at the very moment when Shakespeare was writing *Hamlet*, 'coil' was also coming to mean an altogether neater arrangement of stacked loops. The word seems perfectly suited to describing the chaotic architecture of the human intestines (Hamlet has a preoccupation with the guts, as we have seen), and, more broadly, to communicating a sense of life as a tangled journey both with a beginning and an end and yet also with a cyclical, repeating aspect. Anachronistically, it cannot help but suggest, too, the doubled helical coils of DNA, the molecule of life.

Falstaff's distinctive physical characteristic is, of course, his fatness. He is the 'fat knight', a 'fat rogue', and, more satisfyingly abusive, 'whoreson round man'. To be fat, as Prince Henry scorns, is to sit around being lazy and useless. It is left to Falstaff bitterly to point out that fat has its uses. Aren't fat cattle preferable to the 'Pharaoh's lean kine', he asks. And what of human fat? Towards the end of *The Merry Wives*, Falstaff complains of all the deceptions to which he has been subjected by adversaries who might 'melt me out of my fat drop by drop and liquor fishermen's boots with me'. In Shakespeare's time, human fat was rendered from the bodies of executed and dissected criminals. Called 'oil of man', it was used at least until the end of the eighteenth century as an ointment for wasted limbs – and doubtless, on occasion, too, for waterproofing boots.

I'm reminded of the problematic nature of fat when I see an unusual wax model of a dissected body at the Boerhaave Museum. In the early nineteenth century, Petrus Koning, a morbid anatomist at Utrecht University, took the unusual step of making realistic wax models rather than the idealized versions typically crafted in Italy. Wax models were made as durable substitutes for cadavers for the instruction of medical students. Realism was the ostensible aim of their gifted creators, but some tidying up clearly went on, and sometimes an allegorical dimension crept into the works. They remain more beautiful and affecting than the brightly coloured plastic kits that have superseded them. Koning's departure was to extend the realism of his models to include the yellow layers of fat that other artists chose to omit – and that continue to be left out of today's plastic models.

Our attitude towards our own fat is highly ambivalent. In Genesis, Pharaoh promises that his followers shall eat 'the fat of the land', and it is clear that the fat is the best of what the land can offer. In times when few people could aspire to being fat, fatness was an ideal in itself, a sign of prosperity and health. Wanting for nothing, rulers from Hapshepsut to William the Conqueror to Henry VIII all attained a fine girth. Extreme obesity was not unheard of either. The Greek physician Galen attended one Nichomachus of Smyrna, who was apparently unable to move from his bed for sheer weight. What the philosopher Susan Bordo calls the 'tyranny of slenderness' only began to assert itself in the late Victorian period among a few of the wealthy who reacted ascetically to the greater abundance of food. New scientific ideas about diet and exercise helped them in this direction, as, surely, did the invention of the bathroom scales around this time. It is telling that the latest versions of such scales send small electric currents through your body so as to track changes in the weight of your fat quite separately from the weight of your body overall.

One response to the growing fashion for thinness was to find new labels to suit those who were not thin. The splendid adjective Rubenesque dates from 1913. Derived from the voluptuous nudes painted by Peter Paul Rubens three centuries before, it is a reminder that 'fleshy' was not always bad. A Rubenesque figure is a paradox in the world where size zero is queen. It means not fat, not even large, but somewhat plump and definitely curvaceous – and desirable, not repulsive. It suggests soft flesh, not the hard body of *homo clausus*: more Marilyn Monroe than late-period Madonna.

Scientists interested in attraction have recently examined the Flemish artist's works in order to test the supposition widely accepted by evolutionary psychologists that men are biologically conditioned to prefer women with low waist-to-hip ratios, around 0.70, equivalent to a twenty-five-inch waist and thirty-six-inch hips, despite evidence that in some non-Westernized societies attraction is primarily driven by a high body weight. They measured the waist-to-hip ratios of twenty-nine Rubens nudes, accepted as representing an artistic standard of beauty, and found that they came out substantially higher, at 0.78, providing further evidence to suggest that the hourglass figure represented by the 0.70 ratio is not an ideal for all societies or all times.

So how much fat is too much? Fat is known to serve a number of vital functions, of which the most obvious is the storage of energy. There are around thirty billion fat cells in the body. This figure does not change if you gain weight – at least at first. What happens is that each cell stores more energy-rich lipid, increasing up to fourfold in weight. If weight gain proceeds beyond a certain point, however, these cells begin to divide, and new fat cells are formed. After that, it's harder to lose weight. Fat performs a number of other useful jobs, though, such as providing fatty acids that control cell activity and the hormones that regulate various body functions.

It's clear, then, that fat is not merely stuffing or padding, although it remains far less studied than flesh, bone and the organs of the body. Yet this is how it easily seems to us. It may be present in abundance or largely absent. When it appears, it is amorphous and unruly. It seems continuous, homogeneous, without structure and potentially endless. It serves no visible purpose but accretes anyway, making its own space in the expanding envelope of the body. It flouts the rules of ideal human form. And, bulging out almost anywhere, it makes a mockery of pert *homo clausus*.

Pliny the Elder was one of the first to take against fat, in his *Natural History*, where he tells us that fat is without sense. Flesh can feel and touch, but a layer of fat is a spongy obstacle to sensation and therefore a hindrance to our connection with the world. In contemporary discourse, too, fat is seen not as the necessary complement of flesh, but in some ways as its inverse. Some people even equate the removal of fat with the addition of muscle tissue or the development of a leaner physique. One cosmetic surgeon I interviewed explained that an increasingly popular operation for men is the removal of subcutaneous fat from the stomach in narrow furrows so as to leave the illusion of a 'six-pack', that is to say, a well-developed rectus abdominis, the large, flat muscle that stretches across the stomach, and in which, with good definition, three transverse tendon lines appear.

The questions don't go away when the fat is no longer part of us. Is it waste or a useful resource? It is classified as *clinical* waste in operating theatres. But, surely, it cannot be *bodily* waste if it does not leave the body through the orifices reserved for this purpose. If it has to be cut away or sucked out by force, it must surely be precious to us, more akin to our flesh. Yet, once it is removed, nobody wants it, and the bulky substance inspires some of the same revulsion as normal bodily wastes.

Elaborate social rules and taboos govern our views about blood and excrement. But no such rules have been established with regard to fat. All this uncertainty excites artists. The German Joseph Beuys was renowned for his use of fat in his works. Although he actually used animal fat, it seems clear that we should read the fat as our own. Explaining his use of the material, Beuys made the exotic claim that, when he was shot down over the Crimea in 1944 while serving in the Luftwaffe, he was nursed back to health by Tatar tribesmen who wrapped him in layers of fat and felt.

The increasing popularity of liposuction forces us to confront the cultural meaning of fat anew. What does it mean to remove it? And what does its separate existence outside the body represent? In Ancient Greece, human fat was used for sacrificial and burial offerings, its liquidity thought to contain the essence of life and to counter the dryness of the bones. Today, the rituals take bizarre new forms and suggest new meanings. In 2005, the Australian installation artist Stelarc and his partner Nina Sellars both underwent liposuction and then mixed the fat extracted from their bodies in a large transparent chamber to create an artwork that they called *Blender*. Every few minutes, the mixture is stirred with an electric blender in order to maintain its homogenized liquid state. A major part of the artists' achievement, they say, was to obtain legal ownership of their own bodily residues in order to make the work in the first place. The food writer and self-styled 'gastronaut' Stefan Gates arguably went further when he converted fat extracted from his body by liposuction into glycerol for use in icing a cake, which he then proceeded to eat. What do these stunts tell us, the one a weird simulation of sexual union, the other a shocking exercise in autocannibalism? Perhaps only that our relationship with fat is set to continue on its troubled path.

Tabloid tales of murderers rendering the fat from their victims and selling it for large sums to cosmetics manufacturers, or of plastic surgeons running their cars on diesel fuel derived from their patients' fat, should be treated with caution, however. For most practical purposes, fat from one source is pretty much as good as fat from another, and it is one of the cheapest of animal commodities. There would be no point in going to the trouble of using human fat to do a job that could be done just as well by animal or vegetable fat. If 'oil of man' is to have value in the future, it is more likely to be for the cellular matter that it contains than its energy-rich lipids. In 2002, for example, a team led by Patricia Zuk at the University of California at Los Angeles demonstrated that human fat can be a convenient source of stem cells that are capable of being differentiated more readily into muscle, cartilage or bone tissue than adult stem cells harvested from other parts of the body. We may at last have found a reason to love our love handles.

Bones

The skeletons that dangle so prominently in anatomy rooms ancient and modern, in the medical corners of university bookshops and for that matter from the gibbet beg to be admired as pure structure. When we set out the bare bones of an argument we mean to explain its essentials. Our bare bones represent something essential about us. They are also an aesthetic and engineering marvel.

People rushed to see living bones when the opportunity first arose with the advent of X-rays in 1896. The *Neue Freie Presse* of Vienna announced Wilhelm Röntgen's discovery on 5 January in an article illustrated with an X-ray photograph of Frau Röntgen's left hand. Only her bones and wedding ring were visible; flesh had become transparent. Within days, enthusiasts were making X-ray devices for their own entertainment as well as for medical diagnosis, the application that Röntgen had identified from the outset. Such was the amateur

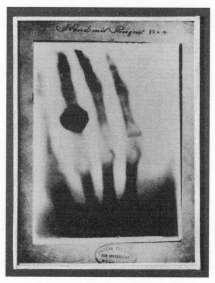

uptake that doctors would ask their patients to bring in X-rays that they had recorded for themselves at home – a practice that led to some nasty radiation burns from the long exposure times required.

The technology makes a notable appearance in Thomas Mann's 1924 novel, *The Magic Mountain*. The naive protagonist Hans Castorp visits an alpine tuberculosis sanatorium. He accompanies his cousin as his X-ray is taken and then proffers his own hand even though he is not ill. He sees what he has been led to expect, 'but which no man was ever intended to see and which he himself had never presumed he would be able to see: he saw his own grave'.

More remarkable is the macabre erotic charge that accompanies these new visions of the body in which one can see not through clothing to the skin but beyond that, through the skin to the bones. At one point, Castorp is shown X-rays of a woman's arm and is reminded: 'That's what they put around you when they make love, you know.' Awaiting his own X-ray, he fantasizes about a woman patient in whom he sees that 'the neck-bone stuck out prominently, and nearly the whole spine was marked out under the close-fitting sweater'.

Immediately, people wanted X-rays to do more than just show them their bones. On 5 February 1896, only a month after their discovery had been announced, the media mogul William Randolph Hearst wired the inventor Thomas Edison to ask if he would record an X-ray of the human brain. Edison seized first the opportunity and then his assistant's head, which he positioned in the path of the X-rays for an hour. But all he could see was the 'curvilinear murkiness' of the man's skull. It had to wait many decades until other techniques would reveal something of this most mysterious body part. X-rays, meanwhile, remain the principal medical means for showing the bones in contrast to the flesh and other soft tissue, and their ghostly negative outlines retain their hold on the public imagination.

There may be sins of the flesh, but the skeleton that carries our weight is the body's innocent slave, a guileless mechanism, admirable in its devotion to duty. It is the only part of the body that endures for ever, which gives it a significance beyond the perishable rest. Although it seems inanimate because it is rigid and hard, the skeleton thus represents the continuation of life (a biologically apt symbolism, as it happens, since bone contains marrow, which generates blood cells).

Chief among the bones of the skeleton in symbolic terms is, of course, the skull. Its eyeless sockets stare, its teeth grin, its lipless mouth accuses. The skull is the ultimate warning against human vanity, the *vanitas* symbol of Classical art, because it is bone but also because it is still recognizable as a face. Such a skull sits ominously on the lid of the tomb in the lesser known of Nicolas Poussin's two paintings entitled *Et in Arcadia Ego* after the inscription that appears on the tomb. As the head often stands for the whole person in life, so the skull stands for the person when dead. A hand-drawn symbol of a skull was once used in ships' logs to record the death of a crew member. This custom probably explains the seventeenth-century origin of the Jolly Roger ensign flown by pirate ships, in which a white skull and crossed bones appear on a black background. 'Jolly Roger' may be a corruption of the French *jolie rouge*, because the flags were once red, an even bloodier indication of the pirates' intentions.

But the skull rejoins its body at other times. Whole skeletons perform the *danse macabre* or *Totentanz*, the dance of death, an allegorical artists' subject that came to prominence in the fifteenth century in the aftermath of the Black Death. When Camille Saint-Saëns wrote his *Danse Macabre* for orchestra in 1874, he made good use of the xylophone to reproduce the spooky clatter of cavorting bones.

Old medical texts typically present the skeleton as one of nature's wonders. The precision engineering of our bones that allows us to walk and run and lift and carry has often been given as evidence for the existence of God. 'I challenge any man to produce, in the joints and pivots of the most complicated or the most flexible machine that was ever contrived, a construction more artificial, or more evidently artificial, than that which is seen in the vertebrae of the *human neck*,' wrote William Paley, in perhaps the most famous of these hymns of praise, his *Natural Theology, or Evidences of the Existence and Attributes of the Deity Collected from the Appearances of Nature* of 1802. Paley admires the vertebrae of the neck in particular because they are articulated so that the head can both nod *and* turn left and right. This is pure teleology, of course: because the bones do their job so miraculously, well, they must have been miraculously formed. Paley it was who came up with the famous analogy of natural creation to a pocket-watch, a mechanism so complex that it could not possibly have been formed without a maker's intervention. The intricacies of human anatomy were a major part of his inspiration.

At any rate, when we look at a skeleton, we see not only an image of mortality but what is clearly some kind of mechanical system. Some of the bones are supporting columns. Others are like beams. They are called upon to work in a variety of ways. Imagine a skeleton holding a bag of shopping. The weight of this load passes up through the bones of the hand and arm to the hinge point of the shoulder. From there, the load is transmitted via the collar bone, shoulder blade and other bones to the spine, where it then runs down through its vertebrae into the pelvis and thence down through the bones of each leg until it reaches the ground. Lifting the bag, the bones of the arm are placed in tension while the spine and leg bones are compressed. The collar bone acts as a beam, with forces of tension stretching the top side of the bone and forces of compression pushing it together underneath as it bends under the load.

We can surely accept without having to invoke a divine creator Paley's view that the bones perform remarkable feats. A slight young woman who weighs, let us say, fifty kilogrammes may possess a skeleton with a dry weight of no more than three or four kilogrammes. This is, I think you will agree, almost incredibly light. It is lighter than some of the plastic replica skeletons that are sold to medical students. Why is this so surprising? In life, we tend to think the bones are heavy and the flesh is light. This is because the latter moves while the former must be moved. We think of muscle as active and bone in contrast as passive and therefore inert and resistant to our will. But I find this perception changes in front of a dissected cadaver. If you've ever held a bone in your hand you'll find that, as soon as you lift a whole limb, it is the flesh that is heavy and the bones that are light.

Dry bones are chiefly composed of hydroxyapatite, a hydrated form of calcium phosphate. The density of this mineral substance is enough to block X-rays. These can then reveal where the bones are inside the body and the material flaws in them, but unfortunately nothing about how they work. They do tell us, though, that there are roughly 206 bones in the human body.

Why *roughly* 206? It's not as if 206 is such a high number that one can't make an exact count. The number is an approximation because certain bones fuse together while we are growing. A bone called the sacrum forms when the lowest five load-bearing vertebrae fuse together where they meet the pelvis. Beneath them, another three, four or five vertebrae fuse together to form the much smaller coccyx, which attaches to the bottom of the sacrum. The coccyx is our vestigial tail. In tailed creatures, many more articulated vertebrae go on to provide its flexible structure. The coccyx might be assumed to be redundant in humans, but it has evolved in parallel with our sedentary lifestyles to serve as the third leg of the bony tripod on which we sit (the other two legs being the catchily named ischial tuberosities of the pelvis): we carry around with us our own bony three-legged stool. Usually, there are more than 206 bones in the body; occasionally, a few more fuse than really should, and we end up with a slightly lower number.

Bones fuse because of gravity. In the effectively weightless environment underwater, the bones of whales and fish may never fuse, and so they carry on growing. Growth is so unimpeded in some cases that size is a good guide to an animal's age. Humans, on the other hand, stop growing at a remarkably constant size. The biologist and philosopher J. B. S. Haldane put it thus in a famous essay:

Let us take the most obvious of possible cases, and consider a giant man sixty feet high – about the height of Giant Pope and Giant Pagan in the illustrated Pilgrim's Progress of my childhood. These monsters were not only ten times as high as Christian, but ten times as wide and ten times as thick, so that their total weight was a thousand times his, or about eighty to ninety tons. Unfortunately the cross sections of their bones were only a hundred times those of Christian, so that every square inch of giant bone had to support ten times the weight borne by a square inch of human bone. As the human thigh-bone breaks under about ten times the human weight, Pope and Pagan would have broken their thighs every time they took a step. This was doubtless why they were sitting down in the picture I remember. But it lessens one's respect for Christian and Jack the Giant Killer.

This argument, incidentally, while persuasive on the ideality of human *size*, all but disproves the existence of an ideal human *proportion*, for if we had evolved to be sixty feet tall, then our proportions would have to have been very different from those given by Polykleitos and Vitruvius.

With 200-plus bones weighing only a few kilogrammes in sum, the average weight of a human bone comes out at less than an ounce. Of course, these bones range widely in both size and shape. The 'longest, largest and strongest bone in the skeleton', as Gray's *Anatomy* puts it, is the thigh bone. With its long, straight shaft and enlarged ball-jointed ends, it makes a handy club, as the apes find in the opening scenes of *2001: A Space Odyssey*. At the other end of the scale are the famously tiny bones of the ear – the malleus, incus and, smallest of all, the stapes (hammer, anvil and stirrup). The stapes, which may weigh as little as three milligrammes, is indeed shaped almost exactly like a riding stirrup.

Many of the names of the bones are visually descriptive like this, even if they sometimes refer to now less than familiar objects. The sternum or breast bone is said to be shaped like a Roman dagger, with its fused parts – the manubrium and the gladiolus – named after its handle and blade. The skull, meanwhile, is likened to a house: the bones at the side are called parietal after the Latin for wall. Below them are the temporal bones, which may relate either to the idea of a temple, as a location on the head appropriate to higher thought, or to the passage of time, as it is here that the hair first starts to turn grey. The clavicle (collar bone), which Gray describes as like an italic letter *f*, gets its name from the Latin for 'little key' (keys were bigger in those days). In the wrist there is the pisiform bone, the size and shape of a pea. Other bones in the hands and feet owe a debt to geometry: the trapezium, trapezoid and cuboid bones. It's all very straightforward if you have conversational Latin and Greek. The question is why more of the bones don't have ordinary Anglo-Saxon names as the major organs and external parts do. Only the spine, ribs and the most obvious bone of all, the skull, have names rooted in the vernacular. Other bones, mainly in the limbs and joints, are named simply after their enfleshed counterparts.

All of these names and descriptions are based on the male skeleton. Feminist reappraisals of historical medical texts have suggested that '[n]o description of the female skeleton existed before the eighteenth century', and there is only one crude illustration from earlier than this, dating from 1605. This lamentable situation has lately been somewhat rectified, but at the price now of regarding the female skeleton largely in terms of its differences from the male, and especially in relation to its supposed function or purpose of child-bearing.

There are in fact many differences between the skeletons of the two sexes, although they are almost entirely differences of degree rather than kind. Women tend to have slimmer bones, a narrower rib cage and a more rounded skull as well as a relatively larger, wider pelvis. (Or, in view of the above omission, we might say that the male may be identified by his heavier limbs, broader chest and more angular skull.) However, the male and female skeletons certainly may not be distinguished by their different number of ribs. The myth that women have thirteen pairs of ribs where men have twelve arises from the biblical story that Eve was created from one of Adam's ribs. Biblical scholars have questioned the meaning of this story. In Hebrew, the word translated as rib (*tsela*) can also mean side, so God's surgery may have been quite major when he fashioned Eve from Adam. This unusual act of creation links Christian theology to other myths, such as that of Dionysus being born out of the thigh of Zeus. Although women do not have that extra rib – thereby making *Spare Rib* a suitably ironic title for the well-known feminist magazine – one in 200 of us does in fact have an extra rib, an evolutionary reminder that we are descended from creatures with many more sets of ribs (such as the serpent in the Garden of Eden, which would have had hundreds of them).

Another apparently obvious difference between the sexes is not reflected in the skeleton. To judge by its name, you would think that the Adam's apple is an exclusively male appurtenance. Yet, as Genesis explains, Eve first tasted the fruit before she tempted Adam to eat from the tree of knowledge. Anatomical fact, both men and women have a feature known as the laryngeal prominence – a bulge of cartilage, not bone – that forms around the larynx. The larynx is a cavity in which air is made to vibrate using the vocal cords. It has a natural resonant frequency governed by the volume of the cavity and the size and shape of its opening. Physicists call such volumes Helmholtz resonators, after the nineteenth-century German physiologist and physicist who made such devices to help him analyse music. An empty bottle is a good example of one. If you blow across the top, it produces a tone at its resonant frequency. If you half-fill the bottle, the pitch of the tone increases. In adolescence, the cartilage around the larynx begins to bulge outward, which increases the volume of the larynx so that it can produce deeper sounds. This bulge is greater in boys than in girls, producing a typical angle of 90 degrees compared to 120 degrees, and it is this difference that explains the greater prominence as well as the deeper voice.

Bone is more advanced in its way than many artificial materials. Bones – including human bones, such as fragments of skull used as scrapers – were among humankind's earliest tools, and today bone continues to inspire materials scientists looking for ways to combine great strength with lightweight performance. As you might expect for a substance that spends most of its time supporting our weight, bone is somewhat stronger in compression than it is in tension. A bone can typically resist a load of a tonne and a half per square centimetre before it breaks. The bones of a child's arm are easily strong enough to support the weight of a family car, for example. Its tensile strength is nearly as great, comparable with that of metals such as copper and cast iron. Only in torsion is bone relatively weak, which explains why most fractures are the consequence of severe twisting forces.

Most bones, especially the long bones of the limbs, tend to be relatively straight. This is not so much in order that they can extend as far as possible for a minimum outlay of material, but because a straight bone has far greater strength than a curved one. The structural columns that support buildings are straight for the same reason. Many of the larger bones are basically tubular in shape. If you cut through them (ask your butcher), you will see an interior structure like a sponge, full of holes. It is clear that this makes the bone lighter than it would be if it were solid. But there is more to it than this. In fact, this is no sponge, but a precisely engineered microstructure providing a network of tiny struts placed just where the bone is most likely to experience forces upon it. Today, furniture designers are beginning to make chairs and tables according to the same minimal principles, using computer-generated force diagrams to tell them where best to place the structural fabric of the object.

It is not any single bone that really inspires; it is how they all work in concert. As the spiritual 'Dem Bones' reminds us (a little incorrectly), each bone is connected to at least one other. To a first approximation, the body is simply an assemblage of straight, rigid beams hinged in various ways at the ends to the next such beam to make up an articulated whole. Few studies were made of the human body as a mechanical system until the launch of the American space programme, when it became important to know how it would respond to the absence of gravity. However, two forerunners in the field were Christian Braune and his student Otto Fischer in Leipzig. Their research during the 1880s arose from early studies of human gait, in turn prompted by the investigations of men such as Etienne-Jules Marey and Eadweard Muybridge into human and animal movement using early methods of high-speed photography. It was a logical extension of this work to want to establish the body's centre of gravity, which Braune and Fischer did by carefully balancing frozen cadavers. They also identified the centres of gravity of major components of the body by cutting them off the cadavers and performing the same balancing tests. Calculations made today – for example, to estimate the extent of whiplash in car accidents – still rely on data from very few original studies like these.

The crude approximation involved in this work hardly does justice to the elegant complexity so admired by Paley. The human skeleton has to perform a huge variety of tasks, including locomotion, balance, resistance and manipulation, all of which expose the bones to high stresses. Normal walking involves fractional adjustments in the position of many individual bones. The gait has half a dozen component actions, for instance, from the pelvic rotation that allows the body to pivot around the stance leg so that the free leg can swing forward until the heel strikes the ground, to subsequent adjustments that transfer the body's weight from the old stance leg to the new forward leg. Many subtle flexions of the knee, ankle and foot ensure that the foot meets and leaves the ground smoothly with each step. The forces that result from all of this complicated activity are equivalent to as much as eight times body weight.

It's all very involved and interdependent. I feel I need to go back to basics, so I turn not to an osteologist but to a structural engineer. Chris Burgoyne is a reader in concrete structures at the University of Cambridge, who has also made studies of the mechanics of bone. Like a proper engineer, he explains things best with the aid of pencil and paper, drawing simple diagrams of lines of force at lightning speed as he speaks. There are three fundamental types of lever, and the body incorporates all three. The first type has the fulcrum – the pivot point – placed between the load to be lifted and a downward applied force, like a seesaw; the other types put the fulcrum at one end of the lever, either with the force at the other end lifting an intermediate load, or with a force in the middle lifting a load at the end. When you lift your finger, you do so using muscles in your arm well above the pivot point of your knuckle. This is the seesaw case: the weight of the finger is on the other side of the fulcrum from the muscular force. Now, use your biceps to lift the length of your arm. This time, the fulcrum is at the shoulder, and the muscle applying the force is positioned between this and the centre of gravity of the arm being lifted. Finally, stand on tiptoe. Now, the upward force is provided by the Achilles tendon and muscles of the leg, the fulcrum is where the toes hinge with the rest of the foot, and the weight of the body falls between the two.

As you will realize from your aching muscles – you may rest now – the bones are not a complete structural framework. They are one complement of what is known as the musculoskeletal system. Any functional structure must have parts that are in tension and parts that are able to withstand compression, otherwise it will either fly apart or crumble. The bones are principally used in compression. It is the muscles that provide the tension. One of Burgoyne's studies involved a structural analysis of the human ribs. The ribs are neither constantly round in section like bars, nor flat like the pieces of a whalebone corset. Instead, they vary in cross-section from trapezoidal near where they join the spine through triangular to elliptical where they terminate at the chest. At first, this seems to make little sense in terms of their function as a protective cage around the body's most important organs. You would simply expect the strongest cross-section to be maintained over the whole length of the bone. However, the ribs are also shaped to accommodate muscle tissue which is attached to them by means of rough ridges on some of the bone surface. This muscle tissue effectively ties the ribs together. When the muscle is taken into account as well, it emerges that the constantly changing rib shape is in fact optimized all along its length for the loads it is likely to experience.

A discussion that begins with mechanical engineering should not omit some consideration of mechanical defects. For the skeleton is not quite so perfectly designed as William Paley and others have thought. The head can nod up and down and turn from side to side, as Paley marvelled, but it cannot, for example, turn through 360 degrees, which might on occasion be rather useful. For all their ability to resist external blows, the ribs are sometimes most at risk from the body itself. A frequent cause of rib fractures is a severe coughing fit when the pressure comes from inside the ribcage.

One surprising advantageous feature of the skeleton is the way that the two main bones of the arm form a rigid rod by using the second bone of the forearm, the ulna, to create a lock at the elbow. By facing the palm of the hand forward, one can then carry a bulky load such as a bucket of water canted out from the body just enough that it avoids banging the knees with each step. In other respects, though, the elbow is of course a weak point, as we are reminded when we bang our funny bone. This point of weakness – where the nerves that run to the two little fingers are squeezed between the elbow and the skin with no muscular protection – is a consequence of our evolution into bipeds. If we still walked on all fours, the forelimb would be angled so that the elbow bent towards the back, not outward, and it would be better protected. Our knees suffer too, as we learn when we reach a certain age, and this too is a consequence of evolution, and our using two feet to bear the weight formerly borne on four. The Achilles heel, however, cannot really be counted as a weak point: anybody would succumb if shot in the heel with a poisoned arrow as Achilles was in legend. This Victorian metaphor is thought to originate instead with Samuel Taylor Coleridge's reference to 'Ireland, that vulnerable heel of the British Achilles'.

The physics is remarkable enough, but bone is also living tissue. It must perform its structural function at the same time as it grows with the rest of the body. Bones develop in response to stress. Tiny cracks form when they are subjected to forces during normal exercise. These cracks send chemical messages instructing new bone tissue to form. However, bone will fail if pushed only a little way beyond its normal performance limit – to about 120 per cent, compared to 200 per cent for materials like steel. 'The body is neither over- nor under-engineered, because all bones have this 120 per cent factor,' Chris tells me. 'It's actually quite natural that you become optimal.' In other words, a bone does not become 'too strong' unless there is some exertion that is making it so, in which case it becomes simply strong *enough*. Conversely, a bone does not usually weaken beyond a safe level unless through lack of use. When sportsmen speak of 'giving it 110 per cent' they are talking more sense than perhaps they realize.

Because of gravity, the body needs to save weight as it grows, as Haldane has explained with reference to the giants in *The Pilgrim's Progress*. It achieves this goal in part by growing bone faster lengthways than across its width (at the price of some reduction in comparative strength in the adult bone). Something clearly guides bone to grow where it is most needed. Whatever this mechanism is – and we will come to it in a moment – it is highly dynamic and responsive to the bodily world around it. It has long been known that bones can be made to increase in size and strength if they are repeatedly stressed. The bones in the spear-carrying arm of a Roman soldier are larger than the bones in the opposite arm, and the same goes for the racquet-wielding arms of tennis players today. Especially in the case of athletic activities taken up in youth, such as ballet or gymnastics, the bones can also be shaped in response to exercise before they harden.

This process allows us to tell much about our ancestors from their surviving bones. We conceitedly believe we are taller than our ancestors because we eat so well. In fact, evidence from skeletons of *Homo erectus* and early *Homo sapiens* shows that they were taller than we are, owing to the strenuous work necessary to survive. From the size of the rough areas on the bones to which muscle attaches, it is known that they were proportionately fitter and heavier too. There is nothing to prevent our regaining these superhuman proportions if only we are prepared to put in the effort – our shrinking stature is not an evolutionary change, but a response to our environment.

Until recently, very little was known about this kind of bone growth. Normal bone growth during development is well understood; it involves the division of cartilage cells on fronts located at the ends of the long bones and their subsequent hardening into bone. But the way that bones respond to use or disuse during life has been something of a puzzle, despite the obvious importance of knowing more about it. Bone can lose up to a third of its mass during the short time a broken leg is in plaster, for example; fortunately, this mass is as quickly replenished when exercise is resumed. They also atrophy in the weightless conditions of space, hence the importance of modelling the behaviour of astronauts' bodies.

Now, to the mystery. Bone exhibits a curious effect known as piezo-electricity. This means that it generates a small electric field when a force is applied to it. This is what happens around the tiny cracks that form when a bone is stressed. Although the details remain unclear, it seems that this effect must be the key to the ability of bones to remodel themselves. New bone cells are created by precursors known as osteoblasts, which carry a positive electric charge owing to the bone-building calcium ions they bring with them. Stress placed on existing bone during exercise generates a negative charge by means of the piezoelectric effect, which then automatically draws these osteoblasts to the site where they are most needed. It is a detail that would have delighted William Paley.

Like Paley, perhaps, we are apt to think of the skeleton as the perfect architectural frame, and so with minor variations it is for most of us. To get a sense of what a wild ride it is to grow a proper skeleton, it is necessary to visit an anatomical collection, such as the one kept at the Royal College of Surgeons in London, which holds the skeleton of a victim of a rare genetic condition called fibrodysplasia ossificans progressiva, in which muscle tissue turns to bone, producing massive calcareous overgrowths that accrete over years, leading to complete immobility. I am forced to acknowledge that the skeleton is not a rock-hard armature like the steel frame of a building, but an entirely organic florescence, subject to shaping by chemicals and external forces.

The ability to grow bone tissue in laboratory conditions is now an attraction to artists. In 2005, for example, Tobie Kerridge, then at the Royal College of Art in London, sought out couples who were interested in a new kind of love token – rings made from the bone tissue of their partner. Potential participants in his Biojewellery Project had to be people who were both due to have their wisdom teeth removed. From small fragments of bone removed during the normal procedure of the tooth extraction, Kerridge was able to culture new bone tissue, which grew and hardened on a ring-shaped scaffold over a period of several weeks, fed by suitable nutrients. Each romantic partner was then able to wear a ring that is a 'part' of the other. 'I cannot imagine anything more intimate, anything more symbolic of our bond, as two individuals, to each other,' wrote one applicant for the project. The couples have various motives for taking part. One pair are materials scientists, another are protesting at the diamond trade, while a third are body piercers taking that art to inner depths. The rings were designed with the participation of the wearers, and carved and decorated in ways that inevitably evoke the 30,000-year human history of working and wearing bones for tools and decoration.

PART TWO

The Parts

Carving Up the Territory

It is the background details in the magnificent illustrations to Vesalius's seven volumes of human anatomy that tell his own story. Engraved on pear wood by an unknown Venetian artist, possibly a pupil of Titian, more than 200 drawings show the body and its parts in all stages of dissection. Vesalius's detailed accompanying text discusses the appearance and functions of these parts, mingling his own discoveries with the respected opinions of Classical scholars and autobiographical anecdotes. When it came out in 1543, *De Humani Corporis Fabrica* became, as its author intended, the most scientifically accurate and complete encyclopedia of the human body ever published – or that would be published for a very long time. The principal subject matter is delineated in a bold and clear manner that is well matched to Vesalius's objective, namely to instruct and enlighten. But there is drama and pathos in the pictures, too. In drawings showing the muscles, for example, flayed skin remains hanging from the body like the drooping clocks of Salvador Dalí. In diagrams of the internal organs, opened torsos have their limbs sharply cut off like the statues of Classical antiquity. The organs heave with visceral realism, but the stumps of these arms and legs are shaded to suggest they are made of sculptor's marble rather than flesh and bone. These illustrations are a perfect fusion of art and science.

The only portrait of Vesalius known to be authentic is contained in one of these woodcuts. It shows him holding a dissected forearm upon which he is demonstrating the working of the hand. (It is certain that both Tulp and Rembrandt were familiar with this likeness.) He is compact and dark-complexioned with wiry, close-cropped hair and an immaculately trimmed beard. His head seems too large for his body, which is certainly diminutive compared with the cadaver he is working on. He turns towards us out of the page and fixes us with a direct glance that has more than a little impishness in it. His attitude chimes with the grim humour in some of the other drawings. In one,

ANDREÆ VESALII.

a muscled body stands knife in hand, triumphantly holding aloft his own excoriated skin. In another, a skeleton man leans nonchalantly on his spade having apparently effected his own disinterment. He gestures with his free arm as if to say, 'Well, what of it?'

But, as I say, it is the little details that are revealing about the man. Hills in the background of this illustration have been identified as those close to Padua, where, in 1537, at the exceptionally young age of twenty-three, Vesalius acceded to the chair of surgery and set about making anatomy central to the curriculum of the most important medical school in Europe. Roman ruins appear in many of the illustrations, perhaps symbolizing Vesalius's demolition of the work of Galen, the Greek physician and anatomist active in Rome during the second century CE, whose writings had dominated medical understanding for nearly 1,400 years.

One engraving is organized so as to present a skeleton in side view. This is the illustration that contains the Hamlet pose, with the skeleton resting its right hand on a skull, which in turn rests on a tomb. The tomb bears the inscription 'VIVITUR INGENIO CAETERA MORTIS ERUNT', an old Latin aphorism which can be taken impersonally to mean 'Genius survives, the rest belongs to death', but might be taken to refer to Vesalius's immodest hope that *his* genius might outlive his rivals'. Behind the skeleton, the stump of a bush is sprigging, indicating that life is both cut off and renewed, a motif that appears in quite a few of the engravings.

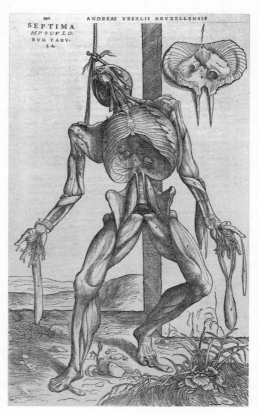

In the first illustration of the volume of the *Fabrica* devoted to the muscles, two putti appear around a decorated initial capital letter in the text. If you look closely, you notice they are no angels, but in fact body-snatchers. Just for fun, it seems, a subsequent illustration chooses to suspend its anatomical subject by a rope, as if it has been hanged, although the rope passes not around the neck but through the eye sockets in such a way as to pull back the head in order to expose the muscles of the throat to the viewer's gaze. The pictures are a reminder not only of the roughness of the times but of the methods to which Vesalius had to resort in order to obtain his research materials. He tells the story of how he stole the remains of a criminal from the gibbet outside Leuven, the Flemish university city where he studied before moving on to Paris and Padua. One day, he goes for a walk 'where the executed criminals are usually placed along the country roads – to the advantage of the students'. He comes across a dried cadaver, its flesh picked away by birds. 'Consequently the bones were entirely bare and held together only by the ligaments so that merely the origins and insertions of the muscles had been preserved.' With the help of a physician friend, he climbs up the post and pulls the thigh bone away from the hip, takes down the shoulder blades with arms and hands attached and brings the parts home 'surreptitiously', making several trips. He leaves behind the head and trunk, which are secured to the gibbet by a chain. But one night soon after, he allows himself to be shut out of the city so that he has time to liberate the rest of the body undisturbed while it is dark. 'So great was my desire to possess those bones that in the middle of the night, alone and in the midst of all those corpses, I climbed the stake with considerable effort and did not hesitate to snatch away that which I so desired.' He conceals the retrieved bones on the ground, and then 'bit by bit' takes them home as well, so that he is able to reassemble the complete skeleton in his bedroom, making up its few missing parts – a hand, a foot and both kneecaps – from other miscellaneous body remains.

Andries van Wesel, his name Latinized as Andreas Vesalius, was born in 1514 in Brussels, the son of an apothecary, and trained in medicine at the University of Paris. Through direct observation, Vesalius largely reinvented the study of human anatomy, modernizing the understanding of Galen, the greatest medical thinker of the Classical period. Galen had brought from Greece the ideas of Aristotle and Hippocrates, the latter still regarded today as the founder of scientific medicine thanks in no small part to Galen's championing of him, and rose in Rome to become the personal physician of the emperor Marcus Aurelius. Galen's concept of the body, formulated in ancient Greece and persisting throughout the Roman Empire and the rise of Christianity, was one of significant parts – chief among them the brain, heart and liver, which respectively governed the bodily compartments of the head, thorax and abdomen. These parts were bound together by the four humours (blood, phlegm, black bile and yellow bile) and a thinner fluid, the spirit, which accounted for the existence of the soul. It was what we might call today a holistic view.

Galen's works had been rediscovered and published in Paris shortly before Vesalius arrived there as a medical student. Vesalius supported this revival, building on it with his own dissections, but also uniquely dared to refute Galen when the evidence of his own eyes did not correspond with the Classical view. An anatomical feature called the *rete mirabile* – the 'wonderful net' – illustrates the shift in understanding. The *rete* is a mesh of veins and arteries found wrapped around the brain in species as varied as sheep and apes. Galen and others believed it was a conduit for the spirit, and Christians subsequently accepted it as the interface between body and soul. Vesalius's early dissections of animals gave him no reason to dispute this, but when he came to perform human dissections while preparing the *Fabrica* in Padua, he could not find the *rete mirabile*, and boldly denied its existence in humans. Vesalius's questioning of Galen was a signal moment not only in the study of anatomy, but also in modern science, implanting the thought that while the Greeks provided a valuable foundation, their ancient knowledge was not unsurpassable. Vesalius was nevertheless careful at first not to alienate his Galenist peers and elders, and it was only in the second edition of the *Fabrica*, which appeared in 1555, that he finally pointed out the error.

We can hardly wonder that medicine made little progress while so many of its ideas of anatomy were based on dissections of animal rather than human parts. Vesalius criticized Galen for this, but was not unknown to take similar shortcuts himself. His plan for the *Fabrica* was to verify everything by reference to the human body, but the shortage of cadavers occasionally drove him to rely on previous published sources or on animal dissections. Although he was the first to describe the prostate gland, Vesalius was generally weak on the reproductive system. His anatomy of the uterus, apparently based on a 'monk's mistress whose body was acquired by dubious means', was reliable enough, but his section on the pregnant anatomy was poor, owing to the paucity of human specimens, pregnant women tending to be healthy women, and his illustration of a human foetus was disgracefully accompanied by a drawing of a canine placenta.

The *Fabrica* revealed the body's interior as a terra incognita ripe for exploration. Anatomical voyagers now set sail to claim its territories, naming body parts like new channels and islands as they went. Vesalius's pupil Falloppio made up for his master's shortcomings by charting the female reproductive system; as we have seen already, the tubes between the uterus and the ovaries, though actually described long before, are now named after him. Eustachi did the same for the ear. Even Nicolaes Tulp got in on the act: Tulp's valve is the puckered portal between the small and large intestines that regulates the passage of digested food waste.

The X-marks-the-spot approach to human anatomy makes a number of dangerous assumptions. It assumes, for example, that identified parts have a distinctive composition or function. This is only sometimes true. It also creates a sense of dividedness, where interconnection may be what matters. Major organs may seem to have a distinct nature and yet are multiply integrated with other parts of the body. 'Bits in between', meanwhile, such as the diaphragm, say, which separates the organs of the chest from those in the abdomen, may be unfairly neglected because they are not seen as forming suitable discrete units.

Carving the body up into parts does have some important advantages. A reductionist approach was essential in order for scientific progress to begin properly to understand the true functions and overturn the old symbolic model of the organs, for example. But it also introduced some troublesome new thoughts. The idea of the human body as a kind of kit that can be taken apart disturbs us because it finds, when all the parts are laid out, that the all-important soul which once seemed to inhabit the body has somehow disappeared. Disassembly also sets up the possibility that a body might be *assembled* – Victor Frankenstein created his monster from body parts (presumably both human and animal) stolen from 'charnel-houses ... dissecting room and the slaughter-house'. Mary Shelley describes the 'miserable monster' very sparingly. 'His limbs were in proportion, and I had selected his features as beautiful,' she has Frankenstein recount. But of course, when it comes to 'bestowing animation', those beautiful parts join to make a horrible, soulless whole.

To investigate how the body actually worked, it was natural to start with the heart, the most dynamic of all the organs, full of moving parts driven by powerful muscles. Body function was fortunately an area where comparison with animals was on safer ground than anatomizing. Vivisection became an important tool. Human vivisection, once practised in ancient Alexandria, was ruled out by the teachings of the Christian church, but there was no restriction on experiments using living animals. If the same organ was observed to respond in a similar way in enough animals of different kinds then this could be taken to be general behaviour that would be found in humans too.

In the mid 1540s, Realdo Colombo, Vesalius's successor in the chair at Padua, gave the first detailed description of pulmonary circulation, the passage of blood via the lungs from one chamber of the heart to the other. (Much later it was learned that Ibn al-Nafis of Damascus had discovered this more than 300 years before.) Vesalius and others had accepted the Galenic belief that blood must pass directly through pores in the muscle wall that separates these chambers, although nobody had observed these pores. Colombo's vivisections showed that spent blood in fact leaves the right chamber of the heart altogether, and travels via an artery to each lung, while veins from the lungs bring fresh blood into the left chamber. Apart from anything else, Colombo's discovery offers a dramatic illustration of the gains that can be made when organs are not regarded as entirely self-contained. Aristotle had held that the blood in the left chamber is cold and that in the right is warm. Colombo was able to correct this. Blood entering the left chamber of the heart is warmer because, as we now know, it has been replenished with oxygen whose reaction with haemoglobin releases heat. He was also able to show that the most important action of the heart is the vigorous contraction that squeezes out the blood and not its subsequent expansion.

Not all of Colombo's experiments were as informative. In one spectacularly tasteless public demonstration, he excised a puppy from the womb of a pregnant bitch, injured it and then showed it to the mother, who licked it solicitously, heedless of her own pain. The scene apparently delighted clergy in the audience as a demonstration of the strength of maternal love even in brute creation.

Colombo's work prepared the ground for William Harvey's discovery of the circulation of the blood around the body as a whole. Harvey – yet another graduate of the Padua school of medicine – was the physician to James I and Charles I of England. In his 1628 book, *The Movement of the Heart and Blood in Animals*, or *De Motu Cordis*, he includes a fulsome dedication to the latter, drawing an analogy between the heart in the body and the king within his kingdom. 'Placed, best of Kings, as you are at the summit of human affairs, you will at all events be able to contemplate together a piece of work principal in the human body and a likeness of your own royal power,' he wrote.

Anatomy, Harvey was accustomed to tell his students, 'informs the Head, guides the hand, and familiarizes the heart of a kind of necessary Inhumanity'. As the historian Ruth Richardson has pointed out, this 'necessary Inhumanity' is what we now call clinical detachment. Harvey certainly had it in abundance. He dissected his own father post mortem and, when his sister died too, completed his knowledge with a complementary dissection of the female anatomy – a stark reflection on the shortage of cadavers available for medical experimentation. In England, a royal annual allowance of four cadavers to surgeons established under Henry VIII was raised to just six by Charles II a century later.

Harvey achieved his breakthrough almost in spite of himself. He was a deeply conservative man who once recommended the diarist John Aubrey go back to sources such as Aristotle and the eleventh-century Persian Avicenna if he wanted to learn medicine, and to ignore trendy 'shitt-breeches' like Vesalius. Fortunately, though, like Vesalius, Harvey believed the evidence of his own eyes. His examination of the beating heart in animal vivisections revealed that its valves work in one direction only, so that oxygen-rich blood returning to the heart from the lungs can only leave again through the aorta, meaning that there must be a circulation of blood around the body equal to that circulating between the heart and lungs discovered by Colombo. Harvey measured the amount of blood pumped by the heart – equivalent to about a double measure of spirits in a pub for every beat. At this rate, the body's full quota of a gallon of blood passes through the heart in no more than a minute! It was clearly impossible that the liver, previously believed responsible for the manufacture of blood, could produce the stuff at such a rate, so Harvey concluded that it must be reused. He pointed to the width of the veins and arteries entering and leaving the heart as further evidence that great volumes of blood must be transported and, holding a beating animal heart in his hand, noted how hard it became each time its powerful muscle contracted.

The discovery of the circulation of the blood caused Harvey some consternation. Others now saw the heart as a mechanical pump in line with René Descartes's conception of the body-as-machine, but Harvey resisted this interpretation. He was more excited by the diagrammatic circularity of the blood's motion through the body, which for him reinforced older ideas of cycles on the cosmic scale. Harvey's discovery was nevertheless of fundamental importance for surgery and all branches of medicine, for example providing new clues to how disease can spread so rapidly through the body, something that had greatly puzzled physicians until then.

On paper at least, the availability of cadavers in Britain should have improved after the passage of an act of Parliament 'for better preventing the horrid crime of murder' in 1752. This stipulated that the bodies of hanged criminals were not to be buried in a normal Christian fashion, but their subsequent dissection could be considered as part of their punishment. Even this step was not enough to sate the demands of the growing medical profession, however. Edinburgh was one city where medical science prospered and the demand for bodies was correspondingly high. In the graveyard of Greyfriars church, you can still see mortsafes, the iron grilles placed over graves in order to prevent the body-snatching that occasioned serious social unrest here and in other British cities throughout the eighteenth century. The body-snatchers, or resurrectionists as they were known, could get good money for fresh cadavers rifled from newly filled graves. (They took care to snatch just the body and not any possessions that might have been buried with it lest they be charged with theft – a dead body belonged to no one, but property still belonged to the relatives.)

In Edinburgh, even the supply of dead bodies from the city's graveyards soon proved inadequate. Between November 1827 and October 1828, William Burke and William Hare, two casual labourers from Ulster, murdered at least sixteen people in the city in order to sell their bodies for the anatomy classes of Dr Robert Knox. It was important that the bodies were not mutilated or damaged, and so Burke and Hare selected victims who would likely be easily subdued. They then plied them with whisky before one of them would place a hand over the nose and mouth while the other lay on top of the body to suppress any struggle. For the best cadavers, Knox gave the men ten pounds.

Knox was not especially well suited to his profession. He found the interior of the human body unpleasant and untidy, lacking in any 'form that sense comprehends, or desires'. He couldn't help seeing his own end in the bodies he cut open. But the external human form, it seems, was another matter. Burke's and Hare's third victim was an eighteen-year-old prostitute called Mary Paterson, whom Knox found so beautiful to behold that he could not bring himself to wield the knife. Instead, in a macabre reversal of the Pygmalion story, he arranged her body into a suitable reclining pose and directed an artist to produce a drawing of her as if living. He preserved her untouched body in whisky for three months more before letting his students demonstrate their skills on her. Many years later, the by then disgraced Knox published *A Manual of Artistic Anatomy*, in which he recalled the perfect body of Mary Paterson, like that of Venus de Milo, with no sign upon its surface to indicate 'the presence of any internal organ or cavity'. This is a revealing definition of human beauty, especially coming from a surgeon, and a telling reaction against scientific reductionism and its body made up of parts.

Edinburgh was not a large city, and the disappearance of familiar figures soon began to be remarked. Burke and Hare were almost exposed when some of Knox's students identified their fifteenth victim on the anatomy table as a mentally retarded lad called James Wilson, well known about town as 'daft Jamie'. Knox denied that it was him, but began the anatomy unusually that day by cutting away the face. The assassins were finally caught when the body of their next and last victim was discovered in their lodgings before they could get it out to Knox's college. Hare escaped the death sentence by giving testimony against Burke, who, in a rich twist of irony, became one of the last murderers in Britain to be sentenced to death and dissection. His skeleton is on display today in the medical museum at the University of Edinburgh.

Burke and Hare are infamous for their deeds, but an earlier episode in British anatomical history is possibly still more gruesome, and goes to the very top of the medical establishment. In 1774, William Hunter published *The Anatomy of the Human Gravid Uterus*, an illustrated atlas of the female reproductive system and the development of the foetus, based on twenty-five years' work and the acquisition of at least fourteen fresh bodies of women who had died during childbirth or in various stages of pregnancy. How did Hunter get hold of these bodies? As we've seen already, heavily pregnant women tend not to fall ill (or to commit hanging offences), so fresh graves and the gallows would offer meagre rewards. As Hunter himself wrote: 'the opportunities for dissecting the human pregnant uterus at leisure, very rarely occur'. Indeed, a major reason for writing the textbook was to give medical students an insight they were unlikely to be able to gain from actual dissections. Nevertheless, it might just be possible

in a crowded city such as London, where Hunter worked, to arrange
to have placed at one's disposal more or less legitimately fourteen
cadavers over a period of more than two decades. In 2010, however,
an art historian named Don Shelton subjected Hunter's *Anatomy* to a
statistical analysis, and came to the conclusion that he must have
worked from the bodies of more women than could conceivably
have been sourced from 'random resurrections', and that some of
them could therefore only have been obtained by a sustained campaign of murder.

The weak link may have been William Hunter's younger brother,
John, who assisted him in his work. John went on to make contributions to many branches of medicine and is acknowledged as the father
of scientific surgery. He may have been the first to use the word
'transplant' in relation to human tissue. His experiments in this field
seem distasteful and misguided today: transplanting tissue from one
part of the same animal to another – moving a spur from the foot to
the comb of a cockerel, for example – as well as transplanting tissue
between animals and between species.

He also experimented by replacing his patients' rotten teeth. Some
human teeth may have been supplied by body-snatchers. But he got
better results using living teeth, and in particular young second teeth
harvested from children. Because second teeth are full-size, and so a
typical boy's tooth would be no smaller than a typical man's, he recommended especially using teeth taken from girls for ease of fit. Even
then there might be trouble achieving a match. In that case, wrote
Hunter, 'The best remedy is to have several people ready, whose
Teeth in appearance are fit; for if the first will not answer, the second
may.' Dental transplantation became widespread until 1785, when it
was established that a young woman had died of syphilis contracted
from an infected implant. As one modern history observes, 'Hunter
seems to have been impervious to ethical criticism.'

There is no doubt that both Hunters would cause some disquiet in
today's medical ethics committees. There is equally no doubt that
they greatly influenced the teaching of obstetrics and helped to
save many infant lives. Their reputation is high today: Glasgow's
Hunterian Museum and Gallery is named after William Hunter,

while the museum of the Royal College of Surgeons in London is named for his brother John. Yet Don Shelton invites us to draw a parallel between them and Nazi scientists such as Josef Mengele, whose data obtained from experiments on Auschwitz concentration camp victims are also available for use by researchers, though frequently shunned. Others have raised similar objections to an anatomical atlas first published in Germany in 1943, which may have made use of bodies harvested from concentration camps. Its author, Eduard Pernkopf, was an enthusiastic Nazi, and some of the artists who produced its coloured illustrations signed their drawings embellished with the SS symbol. As with the Hunters, the moral problem is compounded by the fact that Pernkopf's work is technically excellent, recognized as perhaps the best anatomical atlas since Vesalius. The atlas is still available today in a revised edition with new drawings added since the Nazi period and the offensive SS insignia deleted – except, apparently, in two instances which escaped the publisher's notice.

No name is more closely linked with anatomy in the popular imagination than that of Gray. A few men in time become their books – Webster's Dictionary, Roget's Thesaurus – but rarely one about whom so little is known as Henry Gray. Webster's and Roget are general reference works. So, in their day, were Baedeker's guides and Bradshaw's railway timetables. But why should *everybody* know about Gray's *Anatomy*?

Born in London on an unknown date in 1827, Henry Gray rose without trace. The first records of him show that he entered medical school at St George's Hospital near the family home in Belgravia at the age of fifteen. He did so without having served the usual apprenticeship with an apothecary, suggesting that he had resolved to be a surgeon at a very young age. In a photograph taken by a student friend, he appears with a high forehead and wavy dark hair, somewhat lantern-jawed, with a mouth that dissolves at the corners, Mona Lisa-like. He has dark eyes with a straight gaze and dark eyebrows projecting low over them, giving him a hooded appearance with more than a suggestion of the Romantic poet. Gray quickly showed his mettle by winning important essay prizes – one on comparative anatomy, which was very fashionable at the time, in which he compared the optic nerves of all sorts of (edible) creatures, clearly taking what specimens he could get at the London markets. He followed this with another prize essay, on the spleen, which he converted into his first book in 1854. It was not a success.

Undaunted, however, Gray and his publisher now reached for the broadest possible canvas – the whole human body. Despite its appellation, Gray's *Anatomy* is not the work of Gray alone. It relies heavily on the drawings of one Henry Vandyke Carter, who also entered St George's medical school young, a few years behind Gray. Carter too was studying to be a surgeon, and trying to make a little money on the side by illustrating zoological specimens for the famous naturalist Richard Owen, when Gray approached him to illustrate his new book.

By 1855, when they agreed to work together, both men were qualified surgeons and held teaching positions at the hospital, Gray aged twenty-eight and Carter twenty-four. Their ambition was to produce a new kind of anatomy that would be modern, clear and affordable. The collaboration was necessarily close and prolonged, but not without tension. In the space of less than two years, working at the medical school's dissection room in Knightsbridge, they dissected sufficient bodies to produce 360 illustrations. It is not known how the men obtained the cadavers because the hospital records for the period do not survive, but in general at this time hospitals relied on the bodies of those who died in workhouses and in their own wards. Whatever the truth, their debt goes unacknowledged. According to Ruth Richardson: 'There is a silence at the centre of *Gray's*, as indeed there is in all anatomy books, which relates to the unutterable: a gap which no anatomist appears to address other than by turning away.'

The relationship between the two men is a fascinating one. Carter was aware early on of Gray's rising star within the hospital. At first, the grammar school boy from Hull thought Gray and his set 'snobs', but he soon saw that he was 'very clever and industrious', 'a capital worker' and 'a nice fellow'. Yet when Gray won the Astley Cooper Prize for his essay on the spleen, Carter notes that it was gained at the price of 'beating good men'. He both admires and envies Gray – he complains in his diary of Gray's 'Do it!' approach when he himself cannot get started on things, and feels 'emulous' of Gray's first book, to which he had contributed a few drawings. Yet he also despises him – Gray's aim is 'money', and he is 'not candid quite' in making it clear that Carter will not be credited for his work. When they embark on the much larger, and ostensibly more equal, project of the anatomy, Carter calls the agreement reached for his illustrations 'shabby', but goes ahead anyway. Gray was to earn royalties of £150 for every 1,000 copies sold, Carter just a one-off payment of £150. When Gray saw the proofs of the title page with his name and Carter's set in the same size of type, he struck through Carter's name and instructed that it be set in a smaller size. In later editions, Carter's name was reduced again, and by the seventeenth edition, published in 1909, it was gone altogether.

So Gray's *Anatomy* it was. These are the words that appear on the spine of the first edition of 1858, although the proper title of the work is *Anatomy Descriptive and Surgical*. The publisher's hope was that 'Gray's' single handy volume would be compared favourably with the established multi-volume anatomies – Quain's, Wilson's, and so on. And it was. To the reviewer in the *Lancet*, Gray's *Anatomy* was 'a work of no ordinary labour, and demanded the highest accomplishments both as anatomist and surgeon, for its successful completion. We may say with truth, that there is not a treatise in any language, in which the relations of anatomy and surgery are so clearly and fully shown.'

It was indeed in presenting human anatomy for the needs of modern surgery that Gray's made its lasting reputation. Gray laid his emphasis on what an operating surgeon would actually be likely to see when opening a patient up in order to carry out treatment. With major surgery becoming safer and the recent introduction of inhaled anaesthetics (Queen Victoria received chloroform during the birth of Prince Leopold in 1853), his timing was perfect. His prose is plain and workmanlike, graceless even, with no pretensions to grandeur. Carter's drawings brilliantly fulfil the same brief. They are exceptionally bold – a happy accident arising from the fact that the engraved plates turned out to be over-sized for the book format. Gone are the Classical props and peekaboo antics of Vesalius and other traditional anatomies, Carter being blessedly unaffected by art school training. Instead, text labels appear directly on the drawings in order to help the student link the appearance and name of so many body parts. The art historian Martin Kemp likens Carter's style – or lack of it – to engineering drawings. To me, his illustrations resemble an old department store catalogue or geographical features named on an Ordnance Survey map.

Just three years after the *Anatomy* was published, at the home where he still lived with his mother, Henry Gray died of smallpox caught from his nephew. He was thirty-four. His book, of course, lives on, now in its fortieth edition, produced by a team of eighty-five editorial contributors and twelve illustrators in place of Gray and his single hired artist: Gray's *Anatomy* has become *Gray's Anatomy*.

We began this tour of anatomical science with questions about the discrete existence of organs. Nearly 400 years ago, Helkiah Crooke could already write in his *Microcosmographia*, 'The division of parts into principall, and lesse principall, is verie famous, and hath held the Stage now a long time.' Those principal parts were the heart, the liver and the brain. Galen had counted the testicles as principal too, because of their role in procreation, but Crooke does not award them this high status because they are not essential for survival.

But is the body meaningfully divisible into such parts? I may have body parts, but I cannot 'part', which is to say separate, any one part from the rest of my body without drawing blood. Are the parts, as Darwin believed species had to be, 'tolerably well-defined objects'? Does this separation tell us much that is useful about the body, or does it tell us more about anatomists' attitudes in exploring it?

One body part more than any other shows how the human anatomy is still not fully mapped even now. The clitoris seems to have been known, lost, found, lost again and found once more during the course of 2,000 years of medical history.

The difficulty might have been alleviated had there been more female anatomists. There were a few, especially in Italy where some of the universities gave influential positions to women. In the eighteenth century, Anna Morandi succeeded to her husband's chair in anatomy at the University of Bologna. Her fine anatomical wax models were acquired by Catherine the Great of Russia, as were those of her French contemporary, Marie Marguerite Biheron, who taught John Hunter in London. A century later, Marie-Geneviève-Charlotte Thiroux d'Arconville studied anatomy in Paris, and translated an osteology textbook of Alexander Monro, the founder of a dynasty of Scottish anatomists. She also supervised the drawings for it, while taking steps to preserve her anonymity in the work. She ensured the inclusion of a female skeleton – regarded as highly optional in anatomies of the day – but regrettably allowed culture rather than biology to dictate its appearance. D'Arconville's illustration has the broader pelvis of women, but a disproportionately small head and a sharply tapered ribcage, indicating either that she was unduly influenced by contemporary expectations of ideal feminine form or that she used a model who had spent her formative years in corsets. The d'Arconville skeleton became the pin-up of male osteologists everywhere.

The clitoris was known to the Greeks, regarded either as an imperfect version of the male penis or, in a fanciful analogy with the uvula and the throat, as the guardian placed at the entry of the uterus. This understanding then seems to have been lost in translations of the medical literature from Greek to Arabic and Arabic to Latin during the post-Roman and medieval periods. Falloppio rediscovered the clitoris as a definite body part in the sixteenth century, although it was Falloppio's rival Colombo who published first, adding his own significant observation about its role in generating sexual pleasure. Vesalius, however, was unimpressed. He told Falloppio: 'you can hardly ascribe this new and useless part, as if it were an organ, to healthy women.' He insisted that the clitoris was no more than a pathological feature found only in 'women hermaphrodites'.

The clitoris disappeared from much of the anatomical literature again during the nineteenth century – labelling deleted, for example, from some American editions of Gray – owing to social (male) discomfort about female sexuality. According to Helen O'Connell, an Australian urologist, the worst offender was Last's *Anatomy*, a present-day textbook popular with students cramming for examinations. In other medical textbooks, it is still often adumbrated as the 'female equivalent of the penis', and given cursory schematic treatment perhaps showing no more than the external appearance. A sectional drawing, if included at all, is likely to be one through the mid-plane of the body from front to back of the sort that suffices to show the centrally located functional attributes of the penis, but which does not fully represent the three-dimensional internal extent of the female organ.

The more recent 'discovery' and description of the 'G spot' reveals similar difficulties. In the decadent Berlin of Marlene Dietrich and Kurt Weill, a gynaecologist by the name of Ernst Gräfenberg made his reputation with the invention of one of the first intrauterine contraceptive devices. He escaped from the Nazis in 1940 and eventually set up a private practice in New York, where he was able to pursue his researches into the female orgasm. He did not live to see the term 'G spot' coined in his honour in 1980, and, more to the point, did not in this work himself ever refer to any kind of 'spot', only to a 'zone', involved in female ejaculation. Of course, the G spot is not new except as a cultural construction. Some believe it exists, and others don't, even now.

What the debate sadly shows is that we seem incapable of moving on from the explorer mentality in our investigations of a human body that must, for our convenience, be composed of discrete parts with clear boundaries, like countries, and precise spots where important physiological events are concentrated, like their capital cities. In this sense, we have turned the physical geography of the body into a political geography.

The Head

At the clowns' church, Holy Trinity in Dalston, north London, I meet Mattie Faint, who oversees his profession's dwindling register of members. He is a professional clown himself though dressed in mufti today. The register is not a paper list, but a collection of eggs. There are dozens of them lined up in a cabinet on the wall in a dedicated area of the church. Each egg has been painted, usually in black, red and white to resemble a particular clown, and many have a little felt or conical papier mâché hat on them. A few have a protuberant nose, stuck on like a redcurrant. Some convey signs of the performer's natural appearance beneath the trademark slap – with painted-on crows' feet or facial creases. I look for famous names in the ranks, and locate Grimaldi – a white face with big, friendly eyes. He has big red triangles for cheeks and three tufts of orange hair. Mattie's favourite is Lou Jacobs, who introduced the laughably under-size clown car to the circus ring and whose most distinctive feature is the eyebrows that arch across his face like the McDonald's sign.

The Clown Egg Register is not a joke. It has a serious purpose as an official roster of practising clowns. To be a clown you have to wear make-up – otherwise you're just a stand-up comedian. Traditionally, clowns keep their personal make-up even though they may refresh their act. 'As a clown, unlike actors, you're looked at for who you are,' Mattie explains. 'You're not waiting for a role.' So if you're a clown, your egg is indeed a record of professional identity. Once or twice, an egg has even been exhibited in court to resolve cases of infringement. I find it a happy variant on the unsmiling mug shots by which the rest of us are obliged to identify ourselves.

The register works because we are accustomed both to accept a representation of the head as a sign for the actual head and to accept that the head may stand for the whole person. In Greek and other ancient traditions, the chest is the seat of consciousness, but the head contains the psyche, the principle of life and soul, and the power of the person. A nod of the head is to be heeded as a physical sign of transmission of this power into the world. In its more emphatic, explosive way, a sneeze was regarded as even more significant for being involuntary. It had a prophetic force: whatever the sneezer was thinking at that instant would be fulfilled. This belief held up until the seventeenth century. It is in compensation for the forced expulsion of part of the soul from the head that today we still say 'bless you' when somebody sneezes. In language, we speak of a head of state or of counting heads (perhaps in a poll, a word that originally meant the back of a head). We abbreviate the whole to the head on coins, in sculpted busts and painted portraits, and above all in official identity documents. Signatures, fingerprints, iris scans and DNA profiles may all be used to establish our identity. These ciphers may in future be joined by such biometric arcana as hand geometry, ear shape and skin reflectance, or our personally distinctive voice, gait and key-stroke habits. But it is the facial photograph that has shown itself to be the most broadly acceptable means of official recognition. Any record of identity is always an unsatisfactory – and often somewhat insulting – reduction of our complex self. But the photograph excites less controversy than methods involving a high degree of technological abstraction, because in this case even we can see it is us. It is nevertheless a particular version of you that the authorities wish to see. United Kingdom passport instructions stipulate that you must have 'a neutral expression and your mouth closed (no grinning, frowning or raised eyebrows)'. In other words, no clowning.

The head stands for the whole person never more clearly than when it is set upon a spike. This is the ultimate sign that the body is no more. In death, the head becomes the victor's trophy and a deterrent to others. The weathered head of Oliver Cromwell famously stood outside Westminster Hall for more than twenty years as a warning to would-be republicans, until the spike supporting it broke in a storm, whereupon it was taken into safe keeping by a succession of self-appointed guardians. It was finally buried in the grounds of his old college in Cambridge, the city he had served as a Member of Parliament, 300 years after his posthumous execution, in 1960.

The head is sometimes kept because it provides solid proof of identity, but also for the more superstitious reason that it was thought even in death to harbour the soul. This proposition received an unlikely test during the execution by guillotine of a condemned criminal, Henri Languille, in Orléans on 28 June 1905. A curious physician, Gabriel Beaurieux, examined the man's head as it fell from the guillotine. First, Languille's eyelids and lips went into spasm for five or six seconds, which is a commonly observed reaction. Beaurieux continued to observe, and after a few seconds the man's face relaxed and the eyes turned up. The doctor then did an extraordinary thing: he called out the man's name. He saw the eyelids lift, and Languille's eyes 'fixed themselves on mine and the pupils focused themselves'. As the eyes closed once more, Beaurieux repeated his call, and once more got the same response. 'I was not, then, dealing with the sort of vague dull look without any expression, that can be observed any day in dying people to whom one speaks: I was dealing with undeniably living eyes which were looking at me.' Current medical understanding is that a severed head can remain aware and conscious until falling blood pressure and lack of oxygen causes the brain to shut down, which may indeed take quite a few seconds.

In the Victorian chaos of the Pitt Rivers Museum in Oxford, I find several human heads preserved by shrinking in a display case labelled 'The Treatment of Dead Enemies'. To mitigate my shock, a caption coolly reminds me that taking the heads of one's enemies has been 'a socially approved form of violence' in many cultures, including our own. These particular shrunken heads, or *tsantsas*, are about the size of a cricket ball and somewhat of the same appearance – hard, leathery and mysteriously darkened by age. Some have abundant human hair, some are embellished with streamers. The heads were made by the Shuar tribe of the Upper Amazon in Ecuador and Peru. The Shuar believe that bodies exist in limited numbers. For them, the captured head of an enemy symbolizes the acquisition of an extra body for occupation by your own descendants. When the enemy was closely related by blood, however, it was customary not to take their heads as trophies, but to prepare substitute heads using animals. The Pitt Rivers collection is supplemented with a number of heads of suitably anthropomorphic creatures, such as monkeys and sloths. Like Europeans, the Shuar believe that part of the soul resides in the head, and part of the purpose of shrinking an enemy's head is to pacify that soul.

You may like to know how to prepare a *tsantsa*. First, carefully remove the skin from the skull by cutting a slit upward from the nape of the neck. Discard the skull, brain and other interior matter. Sew up the slit you have made in the skin, and stitch up the eyes and mouth, ensuring that the facial shape is preserved. Then boil the skin until it has reduced to about a third of its initial size. Scrape off any flesh still adhering to the inside. Then cure the skin by repeatedly filling the head with hot pebbles. This will dry it out while preserving its overall shape and characteristic features. The shrunken head is finally suspended by threads. It may then be subjected to verbal abuse, after which its mouth is skewered shut with wooden pins before it can reply.

The preparation of the heads in this manner was a protracted ritual, conducted in stages during retreat from battle raids. Each stage of the process was significant, and the proper enactment of the whole ritual was more important than the finished artefact. The Pitt Rivers Museum holds a number of shrunken heads that it considers to be counterfeit because of irregularities in the way they have been prepared. Today, I am dismayed to learn, the people of the region make heads for the tourist trade stitched together using animal leather.

As the head may stand for the whole person, so the nose sometimes stands for the head. A red nose is enough to advertise a clown, after all. The nose is not the most important facial feature, but it is unquestionably the most prominent, owing to its singular nature, its central position and its forward projection from the head. For all these reasons, the nose gets noticed. Unsurprisingly, therefore, it is also the facial feature about which people tend to be most self-critical. Statistics from the British Association of Aesthetic Plastic Surgeons show that more people undergo rhinoplasty than any other change to their facial appearance. (Coming in a distant second are ear corrections for men and eyebrow lifts for women.)

Nikolai Gogol's hilarious short story 'The Nose', published in 1836, plays on the confusion that ensues when a nose takes on the life of a person. The story begins one morning in St Petersburg when the barber Ivan Yakovlevich finds a nose in his breakfast roll, and recognizes it as belonging to one of his clients, Collegiate Assessor Kovalyov, whom he shaves twice a week. Meanwhile, Kovalyov awakens to find himself with a smooth expanse of skin in place of his 'not unbecoming nose of moderate proportions'. As he goes about his morning chores with a handkerchief clutched to his face wondering what to do, he suddenly passes a 'gentleman in uniform' who is none other than his own nose. As a collegiate assessor, Kovalyov enjoys a rank in the Russian civil service equivalent to that of a major in the army. And the nose? From its gold braid and cockaded hat, 'it was apparent that he pretended to the rank of state councillor'. Kovalyov plucks up the courage to challenge the nose: 'The point is, you're my very own nose,' he blusters. But the nose corrects him: 'I am a person in my own right.' Indeed, it will have nothing to do with its former owner, who ranks lower in society.

Rebuffed, Kovalyov is at a loss how to go about his life without his nose, especially as he is hoping for promotion and a good marriage. This was not part of his plan. Ironically, to wind up with nothing in Russian is 'to be left with a nose'. But he has been left without a nose: what does *that* mean? It is not as if he had lost a toe, he moans; then he could just tuck the injured foot into a boot and no one would be any the wiser. 'If only I had lost an arm or a leg – it would have been far better; or even my ears – that would have been hard, yet I could have borne it; but without his nose a man is nothing.' He tries to place an advertisement in the newspaper, but the clerk refuses, fearing such an announcement would bring ridicule on his publication, which already stands accused of publishing a lot of nonsense. Kovalyov is indignant: 'this is about my very own nose, which amounts to practically the same thing as myself'.

Eventually, the nose is apprehended. Now it must be reunited with its face. 'And what if it won't stick?' At first, Kovalyov tries to reaffix it himself, but it falls to the table with a thud, like a piece of cork. A doctor warns that restorative surgery might only make matters worse. Then, after a couple of weeks, the nose reappears on Kovalyov's face in circumstances just as inexplicable as when it vanished, and Kovalyov resumes his normal life in high spirits as if nothing had ever happened.

It would be foolish to seek too much meaning in what is essentially a brilliant nonsense story. Gogol gleefully exploits the visible absurdity of the human nose. The ridicule that Kovalyov encounters as he walks around St Petersburg, and that he invites from the reader, is amplified by the thought of this most ridiculous facial appendage. The collegiate assessor's acute status anxiety means that, while he professedly never takes personal offence, he simply will not have his rank or title abused. Once his proboscis is restored, he is newly confident but just as status-conscious. After a conciliatory trip to the barber, he visits a pastry shop where he allows himself the pleasure of casting a 'supercilious glance at two officers, one of whom had a nose no bigger than a waistcoat button'.

The temporary autonomy of Kovalyov's nose also provides a playful rehearsal of some of the ideas in Gogol's uncompleted masterpiece, *Dead Souls*, which revolves around the illicit trading in serfs who 'exist' for tax purposes even though they have died. In such a world, the question of a man's ownership of another person, whole or in bodily part, gains a sharp satirical edge. At the end of 'The Nose', Gogol's narrator teases his reader with the observation that strange things do happen, even in St Petersburg, even things that may be of no benefit to the nation. The detachment of a nose from a face, we are led to understand, is not the oddest experience a Russian citizen could expect to undergo. The story occasioned the first of Gogol's several run-ins with the censors for laying bare the absurdities of the system of rank, privilege and favouritism upon which the state depended. It may not be irrelevant to add that Gogol himself had a great beak of a nose.

Conspicuousness begs significance. The size and shape of noses has always provided material for those looking for meaning and difference as well as material for comedy. Why, asks Rabelais's Gargantua, does Frère Jean have such a handsome hooter? Various theories are advanced: God fashioned it so, 'as a potter fashions his vessels'; or, he got first choice when noses were for sale. Jean himself suggests that it 'rose like dough' in the warmth of his wet-nurse's soft breasts. Gargantua adds the bawdy profanity that 'from the shape of his nose you can judge a man's *I lift up unto Thee*'. Tristram Shandy, the narrator of Laurence Sterne's eighteenth-century masterpiece of that title, also refers sadly to the 'succession of short noses' in his family, observing that his grandfather was limited in his choice of wife 'owing to the brevity of his nose'. In short, you don't have to be Sigmund Freud to see that the nose is a symbol of that other forward body part, the penis. There is a hint of this symbolism in Gogol's story, too. When Kovalyov's nose is restored, he finds that he is invigorated in other ways, too, less interested in marriage now, but quite up for sex.

Noses have also drawn the more serious attentions of men with rulers and protractors. The physician and traveller François Bernier was the first to attempt to classify the human population into races, long before the project got into its stride in the nineteenth century. He made a twelve-year voyage to Egypt, the Middle East and India, and wrote an account of his journey called *Travels in the Mogul Empire*. He returned to the salons of Paris dubbed Bernier Grand Mogol, although when Louis XIV asked him which of the countries he had visited he liked best, he apparently replied: 'Switzerland'. In 1684, he anonymously published his scientific ideas in *A New Division of the Earth by the Different Species or Races which Inhabit It*. He divided the world's peoples into four groups: Lapps, Sub-Saharan Africans, Central and East Asians, and a large remainder group including Europeans along with North Africans, people of the Middle East and South Asia and Native Americans. His classification is noteworthy for largely ignoring skin colour and relying instead on physiognomic features, and in particular the shape of the nose. The nose featured in most systematic anthropometric projects thereafter, its casual role in defining race gaining new scientific respectability. That data is useful today in planning nasal surgery, but it never said much about race. The fact is that real people's noses so often fail to conform to the parameters of their supposed racial type that they were always a useless measure of anything at all. It is perhaps odd that Bernier was not alerted to this early on since, as a student in Paris, he had made friends with Cyrano de Bergerac, whose nose seems to have deserved a category of its own.

From such categorization it is but a short step to impute distinctive character to various nose shapes. Inevitably, the phrenologists and physiognomists who inferred human traits from bumps on the skull and facial features had their nasal counterpart. The eighteenth-century Dutch anatomist Petrus Camper tried to gauge intellect from the slope of the nose, a notion based on the fact that this angle changes from infancy into adulthood. 'The idea of stupidity is associ-ated, even by the vulgar, with the elongation of the snout,' Camper wrote. According to his measurements, Classical busts had the most vertical noses, with modern Europeans, Asians and Africans follow-ing in that order. To the race anthropologists who followed in his wake, Camper's metric implied a ranking of races, although Camper himself stated his belief that both black and white shared the same descent from Adam and Eve.

The American publisher Samuel Wells itemized four nose profiles in his popular phrenological annuals (four being a reminder of earlier schemes that linked facial types with the four humours). His ideas were expanded with unpleasant gusto by John Orlando Roe, an ear, nose and throat surgeon in Rochester, New York. In 1887, Roe pub-lished a paper defining five nose types: Roman (indicative of 'executiveness or strength'), Greek ('refinement'), Jewish ('commer-cialism or desire for gain'), Snub or Pug ('weakness and lack of development'), and Celestial. Roe's anti-Semitism is striking – Wells had characterized the 'Jewish or Syrian nose' more kindly as denot-ing 'shrewdness, insight into character, worldly forecast, and dominant spirit of commercialism'. 'Celestial' was Roe's own add-ition. I have absolutely no idea what shape a celestial nose is, although Google Images helpfully informs me that the actress Carey Mulligan has one. Roe says it has the same unattractive attributes as the snub nose, with the addition of 'inquisitiveness'.

Roe's interest in promoting such a typology is all too clear: his speciality was 'correcting' snub noses, to which end he introduced the innovation of operating from within the nostrils so as to leave no visible scar. In late nineteenth-century America, a snub nose was thought undesirable because it was identified with the nose of the degenerate Irish immigrant. Fifty years later, in Nazi Germany, it was the supposedly large nose of the degenerate Jew that was anathema. The nose is seen according to the prejudices of the times.

Laurence Sterne anticipates much of the nonsense that would flow from the scientific measurement of noses and their subsequent organization into 'types'. Tristram Shandy finds in his father's library a treatise by one (fictional) Prignitz, and quotes with approval his findings that 'the osseous or boney parts of human noses . . . are much nearer alike, than the world imagines', and 'the size and jollity of every individual nose, and by which one nose ranks above another, and bears a higher price, is owing to the cartilagenous and muscular parts of it'. He concludes satirically that 'the excellency of the nose is in a direct arithmetical proportion to the excellency of the wearer's fancy'.

The nose features abundantly among the many idioms we use that are based on parts of the body. We nose around in other people's business or keep our nose clean, we follow our nose or pay through the nose, we put somebody's nose out of joint or cut off our own nose to spite our face, we stick our nose in the air or keep it to the grindstone. But most parts of the body, both external and internal, get their turn. We have a nose for trouble, but a head for business and an eye for detail. We could, for instance, rework Shakespeare's 'seven ages of man' speech from *As You Like It* entirely in terms of body idioms associated with those ages. The infant has skin as smooth as a baby's bottom. In childhood, we cut our teeth and dip our toe in the water. The young man may fall head over heels in love and wear his heart on his sleeve. The soldier goes armed to the teeth and, if he has the stomach for it, fights tooth and nail. The justice may be even-handed or he may put his thumb on the scales. Then, in retirement, we take the weight off our feet until, growing long in the tooth, we are on our last legs. Alternatively, we could proceed from head to toe to characterize the ideal man or woman we met earlier, who might have a stiff upper lip, take it on the chin, speak straight from the shoulder, and always get off on the right foot. His or her less fortunate counterpart might be a misery guts who's all fingers and thumbs and has two left feet.

An idiom is defined as a form of words peculiar to a given language or culture. However, many idioms to do with the body have literal translations in other languages. The French, for example, have direct equivalents of our elbow grease, butterflies in the stomach and fleas in the ear; they too learn things by heart, set tongues wagging, and find that things get on their nerves. Like us, the Italians play footsie (*far piedino*) under the table. Other linguistic pairs are more approximate: a sweet tooth is *une bouche sucrée*, a sugared mouth; we feel something in the gut, whereas Germans feel it in the kidneys (*Das geht mir an die Nieren*). Often, a hypernym or hyponym is used, an alternative that encompasses more, or makes do with less, of a given region of the body. We speak of the long arm of the law; the Czechs merely have long fingers. We fall flat on our face; Germans fall, with more precision, on their nose. The synecdoche is total when a single body part stands for the whole person, as it does when we call somebody a great brain or a helping hand, or a prick or an arsehole. Sometimes, languages wander off round the body in search of new inspiration: something that costs us an arm and a leg will cost a Frenchman the skin off his backside or the eyes in his head; and a rule of thumb becomes *une vue de nez*. The same universal bodily action, meanwhile, such as bearing a child, may generate a multiplicity of idioms: to be wet behind the ears has an exact translation in German, but a French naïf is *encore bleu*, while an Italian still has a drip on the nose. In short, few of these sayings are unique to their language as idioms are supposed to be.

There are some exceptions. The Germans seem to favour internal organs. *Ihm ist eine Laus über die Leber gelaufen* (a louse ran across his liver) means he is in a bad mood. *Der hat einen Spleen*, on the other hand, refers to somebody overly obsessed by something. In Hebrew, a person who is not to be trifled with is one who was 'not made with a finger'. Close friends in Spanish are as nail and flesh (*uña y carne*). And in all languages, these phrases are being added to all the time: we now speak of eye candy, a bad hair day, and the arse end of nowhere. There are a few red herrings in the barrel, too. To kick against the pricks is not a modern vulgarism, as you might suppose, suggesting resistance to the idiots who are keeping you down, but a direct biblical quotation referring to the futility of plough oxen kicking against the sticks used to prod them.

Although a few of these idioms are inventive and entertaining, we notice more their sheer obviousness. The body is our most immediate and familiar source of linguistic inspiration. Its parts and our words for them are, quite simply, to hand, at our fingertips, within our grasp, or at least on the tip of our tongue. These examples haven't sprung from famous pens, although many more imaginative ones have, and have often gone on to find their place in the language, as we saw when looking at the body in Shakespeare's works. They are vernacular concoctions, most of them barely similes, merely slight extensions from casual observation. They are obvious, and yet also irresistible in their obviousness. Body idioms tend to be, as Rabelais, Cervantes and Shakespeare were all happy to repeat, 'as plain as the nose on your face'.

We are all, as it happens, 'hairy as an ape'. Humans have just as much hair as chimpanzees. It is only the fact that ours is finer, shorter and generally paler than the chimpanzee's that leaves us free to call ourselves the naked ape. Nevertheless, we make the most of what we've got. Many species spend so many hours grooming themselves and one another that we should never again complain about the time our partner spends at the hairdresser, yet we are the only creatures to have conceived the idea of hairstyle.

Our hair is cultural as much as natural: nothing dates a period film like its actors' voguish hairstyles. What hair we cut, shave or extract and what hair we allow to grow and how we shape it is our decision, but it is a decision strongly guided both by long-standing cultural traditions and by the short-term vagaries of fashion. This applies to body hair, where fashions for shaving armpits, legs and pubic hair come and go. But it applies most obviously to the displayed hair on our head.

Our body hair, and its odd thickets where the limbs join the trunk of the body, are easily explained as residual fur. But our crowning glory confuses evolutionary biologists. It may be chiefly functional, a layer of thatch to insulate our big brains. Or it may simply be what we all feel it is anyway, an evolutionary extravagance like the peacock's tail that provides a basis for sexual selection. Certainly, this is the spirit in which we generally consider the hair. Even the Protestant reformer Martin Luther, whom one would hardly suspect of making such a remark, declared: 'The hair is the richest ornament of women.'

An abundance of hair indicates strength in the male and beauty in the female – and therefore generative potential in both. Hair acquires great narrative value – think of Samson, Rapunzel, Sinéad O'Connor, Britney Spears – from the fact that it may be cut off and, at length, regrown. Its going and coming is an index of these abstract virtues. So it is usually a mistake for the characters in morality tales to grow too fond of their hair. 'God, when he gave me strength, to shew withal / How slight the gift was, hung it in my hair', wails Samson Agonistes in John Milton's poem.

Abundant hair takes the form of shagginess in men, covering large areas of skin, and sinuous length in women. When hidden, women's hair is equated with chastity. Putting the hair up indicates eligibility for marriage. Long, flowing hair is an indication of wantonness – our guilty culture's imaginative extrapolation from nature's gift of hair at puberty. Botticelli's *Venus*, the Lorelei, Rusalka, Mélisande, Mary Magdalene and La Belle Dame Sans Merci all have long hair. The allegorical figure of Opportunity has a lock of hair falling over her eye. *Cherchez la femme*. A tangle of hair is still more troubling. The hair is a trap, like a spider's web, made to entangle men. Belinda in Alexander Pope's mock-heroic poem 'The Rape of the Lock' has her hair in 'mazy ringlets'. And as Simone de Beauvoir observed of Brigitte Bardot: 'The long voluptuous tresses of Mélisande flow down to her shoulders, but her hair-do is that of a negligent waif.'

Strange things happen to hair when it is cut. This dead and yet undead outgrowth from our bodies becomes both fetish and phobia. Trichophobia, a disgust of loose hairs, for example on clothes or clogging the plughole in the bath, is one of the commonest human dreads. It encapsulates the fear of entanglement, but also the sense that cut hair is abject, like nail clippings, spittle and faeces, because it has parted from the body that produced it. And yet we cherish a lock of a lover's hair, and, increasingly, it seems, even wear other people's hair. The singer Jamelia used to wear hair extensions, in order to transform herself, like a cartoon strip heroine, from 'busy mum of two into my alter ego, Jamelia the pop star', until she went in search of their source for a BBC television documentary. DNA analysis of her extensions led her to India, where she found women's and young children's heads being shaved, ostensibly as part of a religious cere-mony, except that the hair was then kept for sale to Western buyers. Though it has gone global, the trade in hair is a long-established busi-ness. Jo March in *Little Women* and Marty South in *The Woodlanders* are among the characters in fiction who sell their hair, while poor Fantine in *Les Misérables* is forced to sell her two front teeth as well. Jo raises twenty-five dollars, Marty two sovereigns and Fantine forty francs – good money.

The women's various reactions to their sudden loss cover the range of evolutionary theories that seek to explain the presence of hair on our head. Jo, as her mother tactlessly points out when the deed is done, has now lost what we have been told several times before is her 'one beauty'. Jo says it will do her vanity good, she was getting too proud of her hair anyway. All four of the sisters are duly wed (as their creator, Louisa May Alcott, never was). However, Jo secures for her-self not the conventional good-looker, but the stout, foreign, middle-aged Professor Bhaer – the rules of sexual selection redux. Having lost her hair, the peasant girl Marty South also loses her mar-riage prospect, Giles Winterborne, who dies, in true Hardy fashion, of exposure. Ironically, he had earlier responded to the shorn Marty's complaining of headaches by saying that it must be because her head is cold. Fantine, meanwhile, consoles herself that she has at least gained her child's warmth in exchange for her own hair.

The Face

In 1859, while scholars sat down to ponder the implications of Charles Darwin's *Origin of Species*, his indefatigable cousin Francis Galton embarked on a systematic investigation of beauty in the British Isles. The young women of London were the most beautiful, he declared finally, and the women of Aberdeen the ugliest.

How did he arrive at his conclusion? Galton, you will remember, was a man given to measurement. During the course of his long career, he sought ways to measure the number of brushstrokes that it takes to make a painting, the parameters of the perfect pot of tea, and the efficacy of prayer (his survey showed that the clergy lived no longer than other professional classes, but then he never asked what they were praying for). To gather the raw data for what he called his 'Beauty Map', he would tear a handy piece of paper into the shape of a crucifix. Using a needle mounted on a thimble, he would then prick holes in the paper to log the numbers of 'girls I passed in streets or elsewhere as attractive, indifferent, or repellent'. The pinholes for attractive girls went into the top part of the cross, those for the average women into the crossbar, and those for the ugly into the stock of the cross. The advantage of this was that he could easily feel for each part of the paper template in his pocket and record his data unseen and unsuspected by whichever town's female populace he was appraising in so un-Victorian a way. 'Of course this was a purely individual estimate,' Galton conceded in his memoirs. But he stoutly defended his scientific method, claiming it to be 'consistent, judging from the conformity of different attempts in the same population'. The project was never completed; perhaps the prospect of a full survey of British girls was too much even for Galton.

The research was not undertaken simply for fun (or indirectly for gain, as beauty 'surveys' put out by cosmetics manufacturers transparently are). So far as Galton was concerned, his data were of little use unless, like cattle, humans could be improved. Darwin had speculated in *The Origin of Species* about the variation of animals under domestication, and this ignited Galton's interest in variation among the human population. Galton coined the word eugenics to describe this ominous project in 1883, but in a way the basic fantasy whereby the rich, intelligent and fecund would be selected in order to improve the British race had little need of modern science. As Galton noted: 'it is not so very long ago in England that it was thought quite natural that the strongest lance at the tournament should win the fairest or the noblest lady . . . What an extraordinary effect might be produced on our race if its object was to unite in marriage those who possessed the finest and most suitable natures, mental, moral, and physical!'

Before the breeding could start, however, there would have to be an awful lot of measurement. This, of course, was Galton's chief delight, and the innocent reason for his pursuit of beautiful girls. As well as seeking field data on the streets of Britain's towns, Galton also sought to capture the essence of beauty through other forms of analysis. One technique developed by Galton was to use the new technology of photography in an effort to identify common facial characteristics among sample populations. He tried both 'composite photography', layering transparent sheets of facial portraits one on top of another in the hope that the blurry cumulative image would amount to a representative average, and, years later, when this had failed to produce meaningful results, the converse process of 'analytical photography', in which faint transparencies of one person in positive and another in negative could be superimposed so that features common to both were cancelled out leaving visible only their supposedly significant facial differences. Both techniques demanded careful preparation, with photographs of the subjects taken at the same size and in the same attitude to facilitate their comparison. Galton gained access to many groups of people distinguished by their deeds or misdeeds or by fortune of their birth. He itemized some of these: 'American scientific men, Baptist ministers, Bethlem Royal Hospital and Hanwell Asylum patients, Chatham privates, children, criminals, families, Greeks and Romans [apparently considered as a job lot!], Leeds Refuge children, Jews, Napoleon I and Queen Victoria and her family, phthisis patients, robust men, Ph.Ds, Westminster schoolboys'. As it turned out, no

SPECIMENS OF COMPOSITE PORTRAITURE

PERSONAL AND FAMILY,

Alexander the Great From 6 Different Medals.

Two Sisters.

From 6 Members of same Family Male & Female.

HEALTH, DISEASE. CRIMINALITY,

23 Cases. Royal Engineers, 12 Officers, 11 Privates

6 Cases

9 Cases

Tubercular Disease

8 Cases

4 Cases

2 Of the many Criminal Types

CONSUMPTION AND OTHER MALADIES

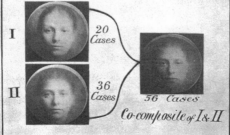

I *20 Cases*

II *36 Cases*

56 Cases Co-composite of I & II

Consumptive Cases.

100 Cases

50 Cases

Not Consumptive.

firm characteristic appearance emerged from the composites. We have to conclude that this list says more about Galton and his times than about any category of individuals.

The main – disappointing – result of all Galton's composite photography was to demonstrate that the more individual subjects were added to the composite image, the more any particular facial characteristic tended to melt away. Even the criminals, in whom Galton was particularly interested to identify a facial type that might be useful to the police, looked quite harmless once a few of their portraits were superimposed.

This tendency had an odd consequence in the case of beauty. As Galton frequently observed, his composite photographs tended to be better looking than the individual portraits from which they were made. The criminals looked less criminal, the sick less unhealthy and so on. The good-looking got even better looking, as Galton found when he photographed portrait busts on casts of ancient coins and medals from the British Museum. In one case, he was thrilled to extract a 'singularly beautiful combination of the faces of six different Roman ladies, forming a charming ideal profile'. The composite in question shows a formidable visage, with a strong, straight nose, a jutting chin and a certain hardness about the lower lip. In quest of beauty, Galton naturally did not neglect the museum's Egyptian coins bearing the head of Cleopatra. He produced a composite photograph based on five specimens: 'Here the composite is as usual better looking than any of the components, none of which, however, give any indication of her reputed beauty; in fact, her features are not only plain, but to an ordinary English taste are simply hideous.'

What does this tell us about the beauty of the human face? Rather than beauty's being in the eye of the beholder, Galton's research invites us to find something objective about it. A composite face, the combined average of several individual faces, is more beautiful than any of the component faces of real people. Yet it is also an average, with all that the term implies. So is beauty simply blandness? Or is it even, perhaps, something more sinister, the human face with the individuality washed out of it? An important talent of fashion models is to be able to look right in different styles of clothes, and for

this a normal face is a good place to start. In 1990, two American
psychologists, Judith Langlois and Lori Roggman at the University
of Texas at Austin, revisited Galton's experiment, using computers
to create superior composite images of women, and this time also
men. By scaling the images so as to superimpose exactly over one
another, they were able to eliminate the blurring that had affected
Galton's composites. They then submitted the resultant images to a
panel of assessors rather than relying on their own personal judgement. Surprisingly perhaps, they confirmed Galton's results. Both
women and men were judged more attractive as composites, and the
more individuals that went into each composite the more attractive it
was judged to be, owing to the progressive elimination of facial
'flaws' and asymmetries. The authors concluded that their findings
were consistent with the pressure of evolution, in other words that
we naturally tend to select partners with characteristics close to the
mean. It's as crushingly unromantic a conclusion as any scientist
could wish for, and Langlois and Roggman seem duly abashed themselves, referring with oblique self-criticism in the abstract of their
paper to science's perennial search for 'a parsimonious answer to the
question of what constitutes beauty'.

Attractiveness turns out to have advantages well beyond the world
of dating. In circumstances where sexual selection couldn't be less
relevant, beauty still has the power to sway our judgement. One typically startling discovery is that attractive persons are more likely to
be acquitted at trial.

Is there more than superficial beauty to be read in the face? If
criminality could be diagnosed from appearance, as Galton hoped to
show, then what of higher virtues? The Greek philosophers believed
that character could be read in the face. The most influential figure in
reviving this idea – called physiognomy – was the Swiss Johann Kaspar Lavater, a Zwinglian pastor, who published a widely translated
collection of essays on the topic in the 1770s. Lavater did his share of
sorting ears and noses into types, believing among other things that
people who looked like certain animals also had something of those
animals' character. 'A beautiful nose,' he suggested, 'will never be
found accompanying an ugly countenance. An ugly person may have

fine eyes, but not a handsome nose.' Lavater himself had a large nose that was almost perfectly triangular in profile, from which we may draw our own conclusions about his self-image.

Above all, Lavater longed to see the face of Christ. The mere sight of it, he believed, would be a divine revelation. It would also provide an ideal template: the more you resembled Him, the better your moral character. The difficulty was that, failing a second coming, there were only artists' representations to go on, and these of course were based on their own ideas of what Christian virtue looked like, perhaps as glimpsed in the faces of virtuous contemporaries. This teleological reasoning tells us nothing in the end about the divine visage, and there is nothing to say that Christ shouldn't look like a wrestler or a truck driver rather than the Californian hippie that artists have settled on.

Like its related field of phrenology, physiognomy is now scientifically discredited. Its chief adherents today are those many authors whose characters' described appearance gives the clue to their behaviour and personality – Charles Dickens's notorious miser Ebenezer Scrooge, of whom we are told: 'The cold within him froze his old features, nipped his pointed nose, shrivelled his cheek, stiffened his gait; made his eyes red, his thin lips blue,' or Gwendolen Harleth in George Eliot's *Daniel Deronda*, whose 'self-complacent' mouth and serpent eyes, described in an opening chapter given over entirely to the arguable matter of her beauty, hint at her later manipulative behaviour, or the unappealing Keith Talent in Martin Amis's *London Fields*, whose eyes shine 'with tremendous accommodations made to money' but without 'enough blood' for murder – and the millions of readers who happily go along with the fiction.

Provoked, perhaps, by Galton's slur on the girls of Aberdeen, Scottish psychologists have been peculiarly active in recent research into our perception of the human face. Computers now allow scientists to manipulate facial images in ways that permit more probing investigation than Galton could ever have achieved with his coins and composites. One particularly striking project, undertaken by Rachel Edwards at the University of St Andrews, involved altering a portrait of Elizabeth I to make it look as if she was using modern cosmetics. Her familiar alabaster foundation – in fact, a poisonous paste of lead white – was replaced by a light tan and an application of blusher. At a stroke, the exercise confirmed the fabled beauty of the virgin queen and provided a convincing demonstration of just how powerful a cultural influence make-up is on our judgement of beauty in appearance.

However, most current studies focus on facial recognition rather than the perception of beauty. Generally, it is more important to be able to recognize a real person than it is to construct an artificial ideal. Galton found this out to his cost one day when he sent some composite photographs he had made of a pair of sisters to their father. 'I am exceedingly obliged for the very curious and interesting composite portraits of my two children,' the father wrote back. 'Knowing the faces so well, it caused me quite a surprise when I opened your letter. I put one of the full faces on the table for the mother to pick up casually. She said, "When did you do this portrait of A? How *like* she is to B! Or *is* it B? I never thought they were so like before."' This was an unusually polite response. Most of his recipients, Galton commented ruefully, 'seldom seem to care much for the result, except as a curiosity'. Galton did not dwell on precisely why people should care about his efforts to make them look more average. But he surely drew the correct conclusion from the rebuffs he received when he added: 'We are all inclined to assert our individuality.'

Identifying an individual, it turns out, is not simply a matter of presenting an accurate likeness. Philip Benson and David Perrett, also at St Andrews, recorded digital images of various faces, and then exaggerated distinctive features to produce a range of more or less extreme caricatures of each one. When they asked people to select the best likeness from the range, the one they tended to pick out was not the true portrait but a slight caricature.

We are actually quite good at identifying faces. It is what psychologists call a natural task. We do it by detecting their overall symmetry, and in particular the inverted triangle described by the eyes and the mouth. The eyes are important because they communicate emotion, while we look to the mouth for signs of pleasure or disgust. This is why we do not notice that Mona Lisa has no eyebrows, or that the *South Park* kids have no noses. Because recognizing faces is so natural, special training, such as that sometimes given to police officers, may actually lead to worse results rather than better, if it disrupts the subconscious image-processing mechanisms at work.

Recalling and identifying a face for ourselves is one thing. But to describe a face so that another person can then identify it as well is a quite different proposition. In these circumstances, we do it by breaking the face down into the bits of it for which we have words. So we start to talk about the eyes, nose, mouth and so on, and perhaps a general head shape (round, almond-shaped, angular, etc.). In addition to functional organs, our conversational inventory of facial components also includes features such as cheekbones, chin and forehead, eyebrows and hairline. The ears may or may not be deemed important according to their prominence (except in Swedish passports, for which it is a requirement that the holder be photographed at an angle so that one ear is visible). Yet this list does not accurately reflect the way we actually identify any face. It is simply the readiest means of communicating what we take to be the characteristic features that make identification possible.

As so often, Leonardo da Vinci may have been the first to assemble an inventory of drawn representations of human facial features, which he did in order to be able to teach fellow artists how to produce recognizable portraits based on only a brief glance at their subject. But in general, from medieval times and through the rise of painted portraits right up until the advent of photography, people were usually identified by means that appear dangerously unreliable to us today, based on signed papers and objects they carried with them, the one set of clothes that they wore, or various distinguishing marks or traits. Sixteenth-century woodcuts of notorious criminals may look like modern wanted posters, but they were invariably produced to broadcast the good news after the criminal had been caught. The idea of producing a likeness in advance, working from people's memories rather than from a living (or by then perhaps dead) subject, did not occur until much later.

In the 1960s, many police forces, looking to improve ways of identifying crime suspects, seized on the fact that we *describe* faces by breaking them down into parts. Early systems such as the American Identikit, using line drawings, and the British Photofit, using photographs, enabled witnesses to piece together an image of a suspect like a jigsaw, using pieces taken from a picture library of stock facial components. The methods worked well in interviews because they relied on a common vocabulary – and seemed more analytical than working with a sketch artist – but produced poor results in terms of likeness. Increasingly, these systems are seen as unsatisfactory because they do not reflect the way we actually *recognize* faces. More powerful computers gradually brought improvements. The E-fit system developed by John Shepherd, a psychologist at the University of Aberdeen, works from a database of complete facial images which may be manipulated and blended according to a witness's instructions to produce a preferred impression. An early triumph of E-fit came in July 1993, when the London serial killer Colin Ireland gave himself up to the police, having realized how closely he resembled his circulated E-fit image.

More recent developments acknowledge the fact that we recognize faces in a holistic way. The Evo-fit system, developed at the University of Stirling, proceeds by showing witnesses a 'line-up' of six real faces from which they pick the best likeness. Of course, it is unlikely that the resemblance will be very close at first, but the procedure is repeated with the best likenesses found at each pass made to 'breed' together to 'evolve' a closer final composite. In effect, the method allows the witness to select for distinguishing facial characteristics while the focus of the exercise is kept on whole faces.

The mistaking of mere resemblance for true identity is a principal cause of miscarriages of justice. This is the true reason behind the Scottish strength in research in this field, which received its initial stimulus from the United Kingdom Home Office following a 1976 government committee investigation into cases where courts had been misled by faulty visual identification. When we say 'his face is printed on my brain' or some such, that may be true as far as it goes. But a slight alteration in that face — a shave, a tan, a haircut, even the same face seen from a new angle — may be enough to confound recognition. In other words, we remember *pictures*, freeze-frame moments, but we know *people*.

Or, we think we do. On 18 October 1997, a boy who had gone missing three years earlier was apparently reunited with his family in Texas, having been found in a youth shelter in Spain. At San Antonio airport, 'Nicholas' was met with hugs and tears from his sister and other relatives. His mother was there too, but held back from the general jubilation. At home over the coming weeks, the boy settled back into normal life, went to school and was able to recall family incidents. If one or two people suspected something was not quite right, the police and immigration officials were on hand to reassure them everything was in order. After a couple of months, however, 'Nicholas' began to unravel. Finally, in March 1998, five months after taking the boy in, the mother communicated her suspicion that he was an impostor, and a cruel deception was exposed. The sixteen-year-old American 'Nicholas' was shown to be Frédéric Bourdin, a twenty-three-year-old Frenchman with bleached hair and a talent for memorizing the details of other people's lives. He was sentenced to six years' imprisonment for perjury and obtaining false documents. On his release, he resumed his career as a serial child impostor, and in 2005 was discovered once more, this time back in France, claiming to be Francisco, a Spanish orphan. The real Nicholas has never been traced.

Society has a desperate requirement that we be in fact exactly who we appear to be. It is not just the Lavaters and the Galtons who want Christ to look virtuous and criminals to look properly vicious. If appearances do not correspond with what we think we know, then the rest of us — in the family, in the community, in authority — are liable to be profoundly unsettled. We may feel affronted, ashamed and threatened when we suddenly learn that somebody in whom we have placed our trust is not what they appear to be. So strong is our need for people to conform to our expectation of them that many conventional statements of identity, including visual likeness, may be overlooked if to do so produces a neater fit. This is what happened to Frédéric / Nicholas. The found boy was too good to be disbelieved; he fulfilled that family's need and tidied up the case-books of the authorities. Such was the social pressure that even the doubting mother was persuaded to accept the impostor.

Personal identity is an act. Most of us settle into one 'character' and maintain it without too much difficulty, in part at least because that is what society requires us to do. Pressure to keep up the act is constant, and we cannot always manage it. So we set aside special times (hen nights, say) and special places (such as on the theatre stage) where we no longer have to *be* who we are; indeed, it becomes socially necessary that we 'are' someone else. In more extreme cases, the balancing act fails with catastrophic results. Bourdin was not able to 'be' himself, so he sought to get along by 'being' other people. Yet the underlying wish is not to pretend, but to belong – for the act to become the life. Unable to sustain the act of his true self, he tried out other acts one after another, but was eventually unable to sustain them as well.

Often what is most astonishing in such stories is not the impersonation, the act performed more or less successfully by the central figure, but the reaction of those around them. As bystanders, we permit ourselves the luxury of being stunned at the apparent credulousness of these people: how on earth can they be duped, we wonder. But from within the story it is clear that their attitude arises from the need to believe, for the sake of personal psychological survival and social cohesion, that the person really is who he or she claims to be. Take the famous story of the return of Martin Guerre. In the mid sixteenth century, a well-to-do peasant by the name of Guerre abruptly leaves his Pyrenean village home and his wife and child with no cause or explanation. Years later, a man returns and is accepted back as Guerre by the wife and by the village. All runs smoothly for a few more years until the wife takes him to court, now claiming that he is an impostor. As the court case is about to be resolved (in the man's favour, you may not be surprised to learn, and besides the rest of the community has no special reason to doubt his identity), the real Martin Guerre makes a dramatic reappearance, minus his leg, as he has been away at the wars.

The psychological core of the story is really Guerre's wife, Bertrande de Rols. Was she a simple woman deceived by the impostor, as other people were, and as (men's) accounts of the episode have suggested? Or was there, as Natalie Zemon Davis, the cultural historian who brought the episode to wider notice, has suggested, reason enough for Bertrande to go along with the pretence? Her status had been much reduced by Guerre's abandonment, and she needed to secure an inheritance for her son. Here suddenly was a plausible and perhaps more satisfactory new mate. 'Beyond a young womanhood with only a brief period of sexuality, beyond a marriage in which her husband understood her little, may have feared her, and surely abandoned her, Bertrande dreamed of a husband and lover who would come back, and be different,' Davis speculates.

Such dramas animate some of the most fundamental and puzzling questions about identity. How do we know we are the same person we were ten minutes or ten years ago? How do our loved ones and others know? Is it even important to be the same and to know it? A Guerre may raise doubts, but when our partner returns from work we are sure it is the same person who left that morning. How we establish this to our satisfaction is no trivial matter. And then, what are we to make of the fact that our body's cells are completely renewed over a period of seven years or so, so that we are materially not the same person at all? The face is our chief frame of recognition, though movement, gesture and voice are important too. Yet the face also changes with age. In what sense do we remain the same really?

Philosophers have always puzzled over what makes somebody a recognizable individual. John Locke and David Hume asserted that consciousness, and continuity of consciousness in memory, were the sine qua non of personal identity. To Hume, 'the principle of individuation is nothing but the invariableness and uninterruptedness of any object, thro' a suppos'd variation of time, by which the mind can trace it in the different periods of its existence, without any break of the view'. Sleep does not constitute such a break because we remember the previous day. But a gap of several years? Well, perhaps that's another matter.

Recent philosophical speculation has tended to focus on hypothetical scenarios in which continuity of self-identity is suddenly disrupted. We are required to imagine, for example, that a person's mind and body are somehow separated from one another and shuffled around in various ways in space and time. But these various thought experiments in time travel, teleportation and body-swapping seem to have added little understanding. Imagining the mind of one person placed into the body of another immediately runs into the problem of the body – is it necessarily of the same gender? Must it feel similar or the same? Imagining a person transported back in time and given the appearance and memories of some historical figure doesn't work either. We can't say that the person then *is* the historical figure because anyone else could have been similarly transported too. The hope of these intellectual exercises is to discover what – since it appears that we have no soul and no other lesser part of us contains our 'self' – really makes us who we are. The experts conclude, rather apologetically I get the sense, that our identity resides in the person most like 'us' a moment ago or a step away – the so-called 'closest continuer theory'. This seems hardly good enough when pushed up against the raw human dramas of taken and mistaken identity.

Neither does it do much to prepare us for the future.

In November 2005, a thirty-eight-year-old Frenchwoman, Isabelle Dinoire, received the world's first partial face transplant at a hospital in Amiens after her dog had chewed off her nose and mouth as she lay unconscious, having taken a drug overdose. More than a dozen similar operations have been performed since then, in France, Spain, China and the United States. The first full facial transplant operation was performed in March 2010 at a hospital in Barcelona on a farmer who had accidentally shot himself in the face. In Britain, a facial transplant team at the Royal Free Hospital in London has ethical clearance to perform four transplant opera-

tions which will be monitored as clinical trials. The first operation
will go ahead as soon as a suitable recipient and donor have been
matched.

I have come to the hospital to talk not to Peter Butler, who will
lead the team of thirty surgeons, anaesthetists and nurses, but to Alex
Clarke, the consultant clinical psychologist who is working with him
on this challenging project. It is her job to prepare potential transplant recipients, but the thrust of her work, given that facial transplant
is still a novelty, has been in helping people come to terms with dis-
figurement rather than dealing with the different set of issues raised
by the prospect of receiving a new face.

Often, it is the rest of us who need helping. 'Societies aren't good
to people who don't look ordinary,' she says. In the past, Alex has
worked with Changing Faces, a charity set up to end what it calls
the 'facial discrimination' felt by people with disfigurements, a
prejudice that is unthinkingly reinforced in popular culture where
the villains are always the ones with scars. Changing Faces has
opposed the panacea of facial transplants, believing that the onus
should be on society to change its attitudes. The Royal College of
Surgeons of England, too, was once against the idea, judging that
the risk of biological rejection of the donated tissue was too high,
but it has modified its position in the light of the Dinoire and other
cases and psychological research, which seems to suggest that the
ethical obstacles are not as great as had been supposed. It is now
cautiously in favour of facial transplantation, while warning of the
dangers of a boom in 'disastrous' operations by 'inexperienced
teams', such as happened following the first successful heart transplants in the 1960s.

The ethics of facial transplantation have no medical parallel. In a
biomedical sense, a face transplant is no different from other transplants – all involve the replacement of diseased or damaged recipient
tissue with healthy donor tissue. But there are some important, and
not necessarily obvious, distinctions of context. The face is external
on the body, normally highly visible, our usual means of human recognition. Along with the hands (once thought to be truer indicators
of personal identity than faces because they cannot change their

expression), our face is the most important representation of our self. Yet Alex tells me this is not a problem. 'No element of identity accounts for squeamishness to do with the hands and face. It's just the newness [of the surgical prospect].'

Alex is more concerned with practical issues. She has found it reassuring to be able to demonstrate using computer graphics that the face that emerges after surgery will not be a macabre mask of the donor's face, but a completely new appearance resulting from the stretching of the donor's skin across the bones of the recipient. 'That helped move us away from the science fiction horror vision,' she says. Almost a bigger concern than the transplant itself is the lifetime of immunosuppressant drugs that will be necessary to ensure that it is not rejected. Potential recipients will have to be screened to assess their likelihood of coping with the physical demands of many operations and the subsequent drug regime. Not all can: Clint Hallam, a New Zealander who lost his hand in a circular-saw accident, had surgery to reattach the hand, which didn't take, and had to be amputated. Some years later, he received the first hand transplant. However, after more than two years using his replacement hand, Hallam voluntarily stopped taking his immunosuppressants, and the hand was once again amputated.

There are also psychological issues surrounding the donor. What kind of person wishes to donate their face? Are they altruists like organ donors, or fantasists who imagine achieving a weird kind of immortality through somebody wearing their face after they are dead? What should the recipient know about the donor's life? Perceptions were not helped when news emerged that Dinoire's donor was a suicide.

Finally, it has to be remembered that, for all its medical glamour, a facial transplant is not, like a heart transplant, in the end a life-saving procedure. Going ahead has to be weighed against alternatives such as skin grafts and other conventional cosmetic surgery or solely psychological treatment. Where the transplant proceeds, there is still a job to explain to the public that its objective is not actually to give the patient a normal-looking face, but mainly to restore important physiological functions such as the ability to work the jaw. There is even some need to check the impulse of people accustomed to the idea of elective cosmetic surgery that they themselves will one day be able to walk out of an operating room with their ideal face. These people may wish to recall Galton's discovery that beauty is only average.

The Brain

Albert Einstein – the greatest scientist of all time, according to many, and the greatest Jew since Jesus, in the words of J. B. S. Haldane – died in the early hours of Sunday 17 April 1955 at his home in Princeton. Dr Thomas Harvey of Princeton Hospital performed the autopsy, and determined the cause of death as a ruptured aortic aneurysm. A dozen of those closest to Einstein attended a brief funeral ceremony. His body was then cremated. A little over fourteen hours had passed since the physicist drew his last breath.

However, not all of Einstein's mortal remains were converted into the ashes that were later scattered at a secret location in order to avoid the attention of celebrity hunters. For at some time during that early Sunday morning, Harvey, acting on his own initiative and without permission from the family, removed Einstein's brain from the skull where it had resided so profitably for seventy-six years and set it aside for examination.

He injected the internal arteries of the brain with formalin and then placed the whole organ in the preserving liquid. The brain revealed no immediate evidence of the special powers that it had possessed when alive. It was carefully measured and photographed, and then cut up into some 240 numbered pieces. Many of these pieces were further sliced into thin sections and encapsulated in layers of a celluloid-like substance so that they could be viewed under a microscope. Harvey appears to have passed many of these specimens out to scientist friends; others he kept. A Chicago doctor reportedly received one specimen as a Christmas present. Another was acquired by a Japanese professor of mathematics who collected Einstein memorabilia. When a journalist tracked Harvey down in Wichita, Kansas, in 1978, he found the remaining chunks of Einstein's brain stored in glass jars in a cardboard box bearing the label of a brand of cider.

Parts of Einstein's brain have been in the hands of scientists for more than fifty years now. What have we learned about how genius manifests itself in the physical body? Harvey promised to publish his findings once he had studied the brain for himself, but for a long time no research was forthcoming. Finally, in 1996, Harvey published a paper in *Neuroscience Letters*, giving the results of his comparison of a prepared section of Einstein's right prefrontal cortex – a part of the brain thought to be involved in governing personality and in judging and comparing thoughts – with those of five elderly control subjects. His earth-shattering news was that Einstein's brain possessed neurons in no greater number and no greater size than the others'.

Marian Diamond at the University of California at Berkeley had only a little more success when she requested a specimen from Harvey and received it in an old mayonnaise jar. In part of the parietal lobe on the top of the head, she found a higher than normal proportion of glial cells to neurons. Glial cells partner neurons in the brain in ways that are as yet poorly understood, contributing to brain growth and function, and are found to increase in animals when they are placed in a stimulating environment. Whether Einstein's glial surplus was present from birth or was the consequence of his immersion at Princeton's Institute of Advanced Studies cannot be told.

Sandra Witelson and others at McMaster University in Hamilton, Ontario, claim to have carried out the first examination of the gross anatomy of Einstein's brain only in 1999. Using calipers, they compared dimensions taken from Harvey's photographs with thirty-five normal male brains, and found no significant differences except in the parietal regions 'important for visuospatial cognition and mathematical thinking'. Einstein's parietal lobes were measured as being about a centimetre wider than the average of Witelson's controls. Unlike all the other male brains – and unlike another fifty-six female brains also examined – Einstein's brain also appeared to be missing a feature known as the parietal operculum, a strip of tissue bordering the lateral sulcus, one of the major clefts that divides the brain into its component lobes. Without this, the Canadian scientists speculated, Einstein's parietal lobes were able to expand beyond the usual size, and to abut more closely with other regions of the brain, with which they may then have built an unusual number of neural connections.

Witelson concludes that 'Einstein's exceptional intellect . . . and his self-described mode of scientific thinking may be related to the atypical anatomy in his inferior parietal lobules', but adds ruefully that her work 'clearly does not resolve the long-standing issue of the neuro-anatomical substrate of intelligence'.

The attempt to locate the origins of genius in great scientists is not new. When Isaac Newton died in 1727, the Flemish sculptor Jan Rysbrack made a plaster of Paris death mask of the great man to assist him in the preparation of the extravagant memorial statue of Newton that now stands in Westminster Abbey. It shows a squat, angular face with a broad, severe crease for a mouth and an equally severe frown. The great man's visage is notably dissimilar to the familiar portrait by Godfrey Kneller, painted in oils, which gives him a long face and red, feminine lips. Rysbrack, too, chose to soften the authentic features for his final sculpture. Both mask and bust have been frequently copied since in plaster, bought and sold like a saint's relics to sit on the desks of pretentious thinkers of following generations. In this way did Newton become one of the favourite subjects of the phrenologists.

The pioneer of this new science, the German Franz Joseph Gall, began collecting such heads in 1792, gradually developing a theory of localization of brain function that he claimed first occurred to him when he was still at school. There he noticed that a fellow pupil with excellent verbal memory had a distinct physical feature: large, protruding eyes. This apparent correlation, observed again in students when he was at university in Vienna, led Gall to believe that an area of the brain directly behind the eyes must be responsible for verbal memory. Gall made systematic measurements of the bumps and dips on the hundreds of specimen heads that he amassed. Believing the best subjects to be those who exhibited the most extreme behaviour or capabilities, he sought out the skulls of murderers, lunatics, renowned statesmen and military leaders, and geniuses in the arts, sciences and philosophy. His ambition was nothing less than to found an anatomy and physiology of the brain that would ultimately reveal a complete psychology of man.

Through his process of 'cranioscopy', Gall identified twenty-seven separate 'organs' of the brain, which he described in terms of the instincts and mental faculties that he supposed they furnished. They included wisdom, kindness, friendship, courage, pride, vanity, caution and firmness of purpose, as well as others to do with senses (of place and space, musical, numerical and mathematical) and the ability to remember (people, words, facts). There were also regions of the brain that he identified with talent as a poet, as a satirist, and as a mimic. At least two of the qualities – the tendency to steal and tendency to murder – surely reflect Gall's access to subjects in prisons. Gall's ideas ran counter to the orthodoxy of the time that the brain was a homogeneous organ whose functions could not be localized, and were deemed too materialistic by the church and by the Viennese authorities. His pinpointing of religious feeling in only one of the twenty-seven areas of the brain may not have helped his case. In 1805, he left Vienna to seek support for his theory elsewhere in Europe, and settled eventually in Paris, where, in 1810, he published his full theory, accompanied by a magnificently illustrated atlas of the brain.

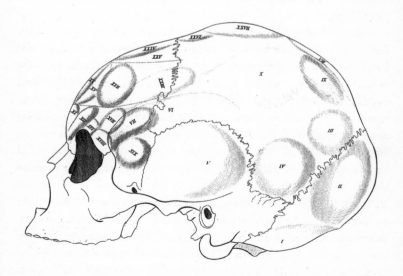

A young British physician, Henry Reeve, was among the many who attended Gall's lectures. Though impressed to begin with, Reeve found Gall vulgar and mannered on a second occasion, writing in his journal that 'like many things seen at a distance, the veil vanishes on narrow inspection and pleasure with it'. Reeve was perhaps not typical, however, as Gall's ideas were to be taken up in Britain, and later in America, with more enthusiasm than they were on the continent.

Gall's chief acolyte on his European tour was his dissectionist and lecturing assistant, Johann Kaspar Spurzheim. However, Spurzheim saw a greater opportunity to popularize the subject that was now becoming known as phrenology, and the two men fell out in 1812. Spurzheim reorganized Gall's system of brain organs, giving many of them appealing new names, and adding a further eight to the list. He divided his now thirty-five organs into the intellectual and the affective or moral. His eye-catching labels included Amativeness, Inhabitiveness, Adhesiveness, Combativeness, Destructiveness, Veneration, Self-esteem, and Marvellousness, which replaced Gall's category of religion. Spurzheim also started the practice of inscribing little desktop busts with these named regions, which proved to be popular souvenir icons of the new science.

Whereas Gall had been concerned with the pure science of the brain, Spurzheim and his followers saw both a moral agenda and a commercial opportunity in demonstrating and offering readings of character. Local phrenological societies sprang up. Having one's bumps felt became the fashionable thing. Every chemist's shop sold phrenological busts. Scientific journals of phrenology began to multiply, filling their pages with detailed analyses of the famous and infamous based on measurement of their skulls or head casts.

The great thing about all this, of course, was that you knew the answer before you began. In 1846, George Combe of Edinburgh, the leading British authority in the field, published a detailed analysis of the skull of the artist Raphael. He described the skull itself as 'a beautiful graceful oval; and its surface was remarkably smooth and equal'. This regularity was central in his view to Raphael's greatness as an artist: 'Taste is the result of a harmonious combination of all the organs, with a fine temperament, and on contemplating these endowments in Raphael, we see the source of his exquisite refinement and grace.' William Stark, the president of the Phrenological Association in Norwich, kept a collection of casts, each of which he captioned with a single personal trait that likewise is curiously consonant with the known facts of the person's life. For example, 'secretiveness' is identified as the most prominent faculty in a man already known to be a 'cunning debtor'.

The great brain of Newton was accorded similar uncritical treatment. In 1845, the *Phrenological Journal* (*and Magazine of Moral Science*, to give it its full title and to distinguish it from various other *Phrenological Journals*) published the findings that Newton's head indicated 'mathematical talents of the highest order', endowed as it was with large faculties of Weight, Form, Size, Order and Number, as well as 'a tolerable share of Causality and Comparison', which explained his ability to trace the relation of cause and effect and to discover analogies, resemblances and differences. The challenge today is to decide whether this kind of thing is actually any less valid than analysis of Einstein's brain that, as we have seen, finds special things to say about the parietal regions at the top of his head 'important for visuospatial cognition and mathematical thinking'.

Phrenology is curious for having serious adherents and all the paraphernalia of proper science – journals, societies, conferences – at the same time as it was attacked by other scientists as quackery, satirized in theatres and magazines, and derided by a sceptical public. For instance, the famous caricaturist George Cruikshank made fun of Spurzheim's faculties of the brain, illustrating 'Adhesiveness' – meaning the propensity to form friendships – by a couple stuck knee-deep in mud. Others suggested bizarre and specific new faculties, such as a talent 'for driving a Tilbury gig'. It is hard to think of another science that has lived such a double life for even a fraction of the century and more during which phrenology persisted.

Phrenology's strength lay in its social promise. The moral dimensions of the human mind could be gauged by the almost ridiculously easy expedient of physical measurements of the head. The phrenologist with an eye to the main chance could set himself up as a proto-psychoanalyst, a careers adviser, a recruitment consultant or even a matchmaker, according to the needs of his client.

Meanwhile, critics of the method noted its obvious flaws, such as the apparently arbitrary number of brain organs, and their interchangeable and contradictory qualities, which allowed any number of conclusions to be drawn from a character assessment. Fittingly, Voltaire's head was adopted as a prop by those seeking to discredit phrenology. Apparently, this 'most celebrated of infidels, and more, the most violent and implacable enemy of Christianity' exhibited an unfeasibly large organ of Veneration. Why did the great philosopher have such a thing if he clearly didn't use it? Phrenology was not to be so easily done down, however. In 1825, one phrenologist wrote that the example of Voltaire actually *confirmed* the technique's veracity, since, while the Frenchman might not show much *godly* veneration, he surely showed it to the French court where he sought his patronage.

As we have seen, Thomas Edison failed to capture an X-ray image of the brain for William Randolph Hearst in 1896. The first rudimentary brain X-rays were not made until 1918, when it was found that air could be introduced into its ventricles to heighten the contrast with the surrounding tissue. However, a practical technique for routinely seeing inside the brain would not emerge until the 1970s. What would it show us? Would it reveal the sites of the powers that lift us above the animals?

Scientific references typically describe the brain as the most complex organ in the human body. It does not look it. It is less multifarious than the heart, less intricate than the lungs. Removed from the head, sliced into sections and squeezed between layers of glass for easy inspection, as I saw it prepared in medical museums, it is white and opaque – literally of course, but also figuratively. It hides its mechanism well. Perhaps it is just human vanity that insists on complexity.

Hippocrates himself may have made the discovery that the brain is not simply a lumpen mass. Around 400 BCE, probably based on his examination of Greek soldiers injured in battle, he compiled a book called *On Injuries of the Head*. Here he noted, for example, that injuries on one side of the brain tend to lead to convulsions on the opposite side of the body. Later, Galen sought the location of the soul in the brain, and made reference to the brain's having parts dedicated to specific body functions. Medieval figures such as the Persian scholar Avicenna regarded the four ventricles of the brain that contain the cerebrospinal fluid as storage spaces for images and ideas, respectively governing perception, imagination, cognition and memory. Much later, Descartes felt he had located the soul in the tiny pineal gland at the base of the brain. The phrenologists did little to advance the science, but they too shared the conviction that the brain was not a homogeneous and holistically functioning unit, but an organ of distinct parts. This conviction has strengthened with the advent of new ways to probe and map the brain.

The methods are often brutal. As in Hippocrates's day, war is a spur to knowledge. In the Russo-Japanese war, an ophthalmologist named Tatsuji Inouye was able to map the visual cortex in new detail based on gunshot wounds received to the occipital lobe at the back of the head. He benefited – if that's the word – from the Russians' use of new guns that fired bullets that were more penetrating but less damaging to the surrounding flesh than previous weapons. British neurologists were similarly able to make strides in understanding the role that the occipital lobe plays in vision because the Brodie helmets worn by British soldiers provided such poor protection in this area. (Unfortunately for them, the phrenologists had tended to locate visual faculties unimaginatively just behind the eye, nowhere near the occipital lobe, to which they ascribed qualities of love and friendship.)

Later studies by the American-born neurosurgeon Wilder Penfield in Montreal observed the response of conscious epilepsy patients to brain stimulation by electrodes. Penfield used the technique in order to plan brain surgery to relieve convulsions experienced in specific regions of the body. But what he obtained as a result was a new map of the brain. Penfield's masterstroke was to employ an artist when he published his findings in 1937. Mrs H. P. Cantlie drew a 'cortical homunculus' in which the various sensory and motor functions of the body were depicted at a scale in proportion to the volume of the area of the brain thought to be responsible for their control. Unfortunately, the diagram – showing greatly enlarged thumbs and large fingers, hands and feet compared to the limbs and trunk of the body – looked a bit like a frog squashed on the road. More instructive and enduring is a later version that Penfield published in which the sensory and motor organs are draped directly around the hemispheres of the brain in a sectional view across the head. The lips and the thumb stand out especially. This graphic idea has taken root and inspired ever more grotesque variants since, as well as having its precursors in the homunculus of medieval belief, who was literally a little man, or 'manikin', a kind of Mini-Me that might be conjured by an alchemist or a magician. These distorted human figures are perhaps also imaginatively invoked in the gangly dragons and monsters of our nightmares and cartoons, with their grasping fingers and clumping feet.

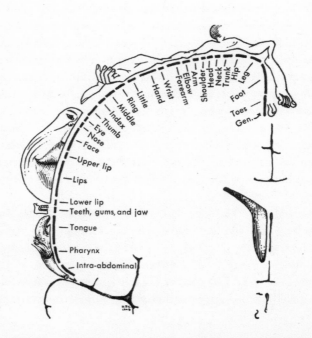

True images of the brain are brought to us by a different magic. The secret is the phenomenon of nuclear magnetic resonance, a discovery of such momentous significance that it has been marked by the award of Nobel Prizes on six occasions: three in physics, two in chemistry, and the latest in medicine, awarded in 2003 for its application in the form of medical imaging now universally known as MRI.

I went to have my brain scanned more than twenty years ago. It was the spring of 1988, and this form of imaging had only just gained approval for clinical use. So new was the technique that nobody had yet thought to ditch the 'nuclear' part of the name that somehow failed to reassure prospective patients. I am not a patient, however, but writing an article for *Popular Science* magazine.

When I arrive at the Albany Medical Center in the New York state capital, the white-coated head of neuroradiology, Gary Wood, begins by asking me some preliminary questions. 'Is there anything wrong with you? Do you have anything metal on you – pens, paper clips?' I deposit keys, a pen, and my tape recorder in a locker. Then the doctor opens a big door shielded with copper and ushers me into the MRI room.

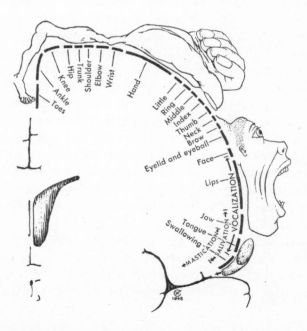

A large doughnut-shaped machine fills the room. It is emblazoned with the logo of General Electric, the company founded nearly 100 years before by Thomas Edison, funnily enough, and based in nearby Schenectady. Smoothly contoured white plastic conceals its five-tonne magnet. (Medical NMR magnets may generate magnetic fields measuring some 15,000 gauss; the Earth's magnetic field by comparison averages just 0.5 gauss, while the magnet in your fridge door might produce around 50 gauss.) Wood's assistant helps me on to a gurney that projects from the bore of the magnet, and then flicks a switch. Powered by hydraulics (motors won't work near this huge magnet), I glide almost silently into the magnet until my head is positioned at its centre. Any sense of claustrophobia is mitigated by the mirror thoughtfully angled above my eyes so that I can see out beyond my feet and through the room's observation window to where Gary and his colleagues are monitoring the scan. Through a two-way audio link, I hear them typing instructions into the computer and chatting excitedly about their new equipment. 'Lie still,' I am told. Gary presses a button. A rapid, dull drumming fills my ears, but I feel nothing as the massive machine scans the depths of my brain.

Afterwards, Gary shows me what he has recorded on the monitor. It is the first time I have been able to see inside my own body. Yet even at this early date, I find I am jaded by the generic familiarity of the images. 'MRI has shifted our sense of transparency so that we can see those structures whose form and function had previously been the domain of poets and philosophers,' I read in one rather awestruck history of medical imaging. But what is seeing? I am aware that what I am looking at is not a simple photograph, but a highly indirect image, a digital manifestation of a series of radio-frequency signals, which are themselves the product of tiny magnetic fields produced by hydrogen atoms in my brain in response to the massive input signal of the imaging machine. It seems to me the poets and philosophers might still have the edge.

Sensing my ambivalence, perhaps, Gary points to different shades of grey on the screen that represent the outer shell of my skull, my bone marrow, and even my cerebrospinal fluid. 'Now we're going to page through,' he tells me. 'We're going to drive right through your head.' A series of images appears on the screen as Gary chases my optic nerves from my eyes into my brain. He pauses at one picture, a cross-section clearly showing my nose, throat and sinuses. 'Here's something that looks like a Dristan commercial,' he laughs. As I depart, he gives me a souvenir print of my scan. Sadly, I no longer have the image, so I cannot tell whether my parietal lobe is expanded or my lateral sulcus closed up like Einstein's.

Improvements in magnetic resonance imaging made since the time of my scan now allow scientists to obtain live, moving images of the working brain. Experiments in *functional* magnetic resonance imaging (fMRI) typically involve scanning a subject's brain while he or she performs particular tasks. This yields images that highlight the parts of the brain that are temporarily more active. The digital image processing applied to the MRI scans generally displays a section through the whole brain in black-and-white with the active area shown as a coloured highlight. Thanks to this manipulation, we now speak happily of the parts of the brain that 'light up' when we think particular thoughts, although, strictly speaking, the observed 'lighting up' is an indication of increased blood flow and not necessarily of a particular mental activity.

This new technology is an important aid in diagnosing brain disease, but it also provides a new tool for investigating the way the brain works normally. Many studies are underway to examine aspects of human mental activity that we tend to regard as important in defining who we are as individuals. These include the making of moral choices, the display of prejudice, and the exercise of personal creativity. Even simple decisions without consequences require the exercise of choice, which is an expression of personality. Neuroscientists at the Oxford Centre for Functional MRI of the Brain devised an experiment that required subjects to push buttons in order to switch from an arbitrary state A to states B or C. When subjects chose freely, their action was accompanied by increased activity in one particular part of the brain and reduced activity in another. When the same subject was directed as to what to do by a second person, however, this picture was reversed. The experiment seems to show that the neural mechanisms underlying our assessment of the choices we make are different according to whether those choices are forced or freely made.

But what about a real moral dilemma? Joshua Greene at Harvard University asked his subjects to imagine a situation in which a crying baby threatens to give away the presence of a group of people hiding from enemy soldiers: do you smother the baby to save the lives of the others? His results showed that brain regions associated with planning, reasoning and attention were comparatively more active when people chose to harm some individuals in order to save others. In other words, people think harder when what they decide will have consequences for others. It is what we would at least hope for of our fellow human beings.

Greene's colleague at Harvard, Jason Mitchell, has been using fMRI to investigate empathy and prejudice. Understanding other people involves imagining ourselves in their position. This is easier to do when the other person is similar to oneself. Mitchell asked his subjects, defined according to their social and political beliefs, to evaluate imagined persons both strongly like and strongly unlike themselves. The brain images he recorded show that the perception of a similar 'other' engages a region of the brain known to be linked to self-referential thought, whereas perception of a dissimilar 'other' activates a different region. It does not reveal *why*, but it does show a little of what happens *when*, for example, white people more readily associate black faces with negative attributes and white faces like their own with positive ones. Such work may provide a key to understanding racial and other forms of prejudice.

Creative works such as paintings, symphonies and novels are seen as highly personally expressive. But can the creative process be seen as it happens in the brain? Charles Limb at the Johns Hopkins School of Medicine in Washington, DC has tried to catch a glimpse of it by recording fMRI scans of skilled jazz musicians as they improvise at the piano – devising music never thought of or played before. An average made of the brain images of a number of improvisers shows particular areas of the brain activated and others deactivated, suggesting that creativity, too, is localized. Imaging studies of the normal

brain such as these gain validity by taking data from a sample population of subjects, not just a single person. I can see how it might be
dangerous to interpret one person's scan in a particular way when
looking at something as personal and subjective as prejudice or creativity. Yet I can't help wondering if these statistical aggregates, like
Galton's composite photographs, risk throwing away the very information they are trying to gather.

Functional MRI is also being applied for less lofty purposes. Brain
scans of people on slimming regimes made as they choose whether to
eat healthy or junk food, for example, appear to highlight areas of
the brain involved in self-control. Product manufacturers and advertising agencies are naturally very interested in this activity of the
brain – and in being able to circumvent it. Now that MRI has proven
itself as a diagnostic technique and the cost of the equipment is falling, businesses are starting to think about what it could offer them.
Gemma Calvert is a former academic psychologist and now managing director of Neurosense, a company that uses brain imaging to
probe the mysterious recesses of the consumer's mind. 'There is a
perception out there that this was developed as a medical technology,
and now you're using it for commercial purposes, and what are you
playing at?' Gemma admits. But major corporations clearly have no
such qualms. Neurosense used brain imaging on behalf of a British
breakfast programme on commercial television, producing the self-
serving result that viewers paid more attention to, and were better
able to recall, advertisements screened in the morning.

'You shouldn't be sceptical that this technology allows us to see
how the brain performs a certain task,' Gemma chides me. 'The
tricky bit comes when you start asking more social questions. Will
you ever really be able to use these technologies to read what I'm
thinking? I for one would like to see that.' The prospect remains theoretical for now, though, requiring scanners with much greater
resolution than those available today that could capture the firing of
individual neurons in the brain. This might indicate what someone is
thinking in response to a certain stimulus. 'But that still doesn't get at
the sense of experience. The sense-of-being-alive thing is a biggie.'

In San Diego, meanwhile, a company called No Lie MRI shows

one direction where this technology is headed. It hopes to use fMRI to enable its clients to assess job applicants and insurance claimants. Because the imaging technique monitors the central nervous system directly, rather than the autonomic nervous system that controls body functions, No Lie MRI claims it is able to bypass American legal restrictions that apply to companies' use of polygraph lie detectors. Its plan is to set up Orwellian-sounding VeraCentres where subjects will be interviewed while being scanned by an MRI machine. The company is presently lobbying so that fMRI 'evidence' will be admissible in American courts. Even neutral organizations such as the British Psychological Society concede that it is probably only a matter of time until brain scans are admitted in court, even though, as with DNA evidence, the aura of science that surrounds them can mean that jurors give them a credence that they do not always merit.

In its zeal to catch fibbers, No Lie MRI may be missing the big picture. To a neuroscientist, and increasingly to all of us, we are our brains. The day may not be far off when a man can walk into court and accuse his own brain of the crime, and the evidence will support his claim. Or, to put it another way, any defendant in future may be able to plead a sophisticated modern equivalent of the insanity plea. The question then is whether it makes any sense to punish the person – or their brain.

The Heart

The heart is a hollow muscular organ of a conical form, placed between the lungs, and enclosed in the cavity of the pericardium.

The heart is pyramidal, or rather turbinated, and somewhat answering to the proportion of a pine kernel.

The heart of creatures is the foundation of life, the Prince of all, the Sun of their microcosm, on which all vegetation does depend, from whence all vigour and strength does flow.

The heart, like a chasuble.

The heart, like a fleshy whoopie cushion.

The heart is deceitful above all things, and desperately wicked.

The heart has its reasons of which reason knows nothing.

The heart is a hungry and restless thing; it will have something to feed upon. If it enjoys nothing from God, it will hunt for something among the creatures, and there it often loses itself as well as its end.

The heart is forever inexperienced.

The heart is a lonely hunter.

The heart, then, is many things to many people, as these varied descriptions attest. The first three descriptions here are by anatomists at different periods, taken respectively from Gray's *Anatomy*, Helkiah Crooke's *Microcosmographia* and William Harvey's *De Motu Cordis*. The next, 'The heart, like a chasuble', is from *Pantagruel* by François Rabelais, who was an anatomist as well as a monk, a lawyer and a writer. On one occasion, in Lyons in 1538, a corpse spoke to Rabelais, at least as told in a contemporary poem by Etienne Dolet. The corpse clearly felt he had got his own back on the judges who had only sought to increase his punishment by sentencing him to death with dissection when he learned he was to be dissected by the great Rabelais: 'Now Fortune you may rage indeed: all blessings I enjoy.' The next, possibly more informative, simile comes from Louisa Young's *The Book of the Heart*. The remaining statements are drawn from the

Old Testament Book of Jeremiah, the seventeenth-century French philosopher Blaise Pascal and his contemporary the English clergyman John Flavel, and the American writers Henry David Thoreau and Carson McCullers.

The idea that the heart represents in some important way our very core goes back to Aristotle and beyond. According to Young, Egyptian and Greek stories of more than 3,000 years ago reveal that the heart was already regarded as the seat of 'identity, life, fertility, loyalty and love'. Whether this was physiologically true was to remain unknown for many centuries. But the fact that it was absolutely the case in a symbolic sense was underwritten for some 1,300 years when Galen in the second century CE placed the liver, heart and brain in charge of the tripartite body (abdomen, thorax and head), the heart inevitably central of the three.

Unlike all the other internal organs, the heart is clearly discernible as a site of activity: it beats, and beats at a rate that changes in response to the world around it, faster in the presence of a lover, or of danger, slower in sleep and at the approach of death. Classical physicians saw the heart as the source of the body's heat and as connected with the blood, but it is astonishing that its true function as a pump sending the blood round the body was not understood for so long. Leonardo da Vinci came tantalizingly close to the truth when he observed, as Galen had not, that the heart has four chambers, is highly muscular, and is the source of all blood vessels. Had he only noticed that some of these vessels carry blood out from the heart and others return it, he surely would have drawn the obvious conclusion, and sealed his reputation as rather more than an amateur in the field of anatomy.

When I hold a heart in my hand, it is immediately obvious that it must once have done something. Compared with the lungs or brain, liver or kidneys, which have an inscrutable uniform texture, this organ has a convoluted architecture. I pull aside the thin folds of fat that wrap it like tissue paper round a piece of china. It has a muscular base with its various chambers (two atria and two ventricles) above. Empty of blood, it is noticeably bottom-heavy. Blood vessels trace wormlike paths across its external surface. This particular heart has been cut away across the aorta, revealing it as a huge tunnel about two centimetres in diameter. I read that the heart pumps 10,000 pints of blood in a day, and that it can squirt blood six feet into the air through this tube. How, in the past, could people imagine that the body simply manufactured blood at the colossal rate that this gaping conduit surely demands? The major vein of the body, the vena cava, is almost as big where it enters the heart. Four other large blood vessels, the pulmonary veins and arteries that transport blood to and from the lungs where it is oxygenated, are about a centimetre across. The whole design reminds me of a diagram of an underground train station. I imagine it will be a puzzle to place the heart back in the prosected body from which I have lifted it so that the severed tubes meet up, but in fact it slips easily back into the hollow left for it by the lungs, finding its correct orientation as it falls into place, like an animal settling in its nest.

In places, I can see wavy flaps of flesh. These are the valves that regulate the blood flow. They create the characteristic double beat of the heart, which is typically spelled out as 'lub-dub' or 'lub-dup'. If you speak these two syllables aloud your tongue will mimic the action of the two sets of valves that regulate the flow of blood. Part of the reason why William Harvey was able to discover the circulation of the blood where Galen and Leonardo had failed may be owing to advances in hydraulic engineering made in the early seventeenth century, including, oddly enough, Pascal's invention of the hydraulic press. Perhaps these water-pumping contraptions enabled Harvey to see the heart afresh. In any case, Harvey elucidated the mechanism of the heart and blood flow with exemplary scientific clarity, although he remained baffled as to what all the activity was for. This would have to await the discovery of oxygen and the role of red blood cells more than a century later. Sadly, once his book was published, things went less well for Harvey. His friend and biographer John Aubrey wrote that he 'fell mightily in his Practize, and that 'twas beleeved by the vulgar that he was crack-brained; and all the Physitians were against his Opinion, and envyed him; many wrote against him'.

Harvey's breakthrough did no harm to the conventional reading
of the heart as the centre of things, though. Though in fact slightly
off-centre (to the left) within the body, the heart represents a sensible
centricity, a midpoint between the head and the sex, the fulcrum of
reason and lust. Its newly understood role in pumping blood round
the entire body simply strengthened its metaphoric importance, as
Harvey himself was not slow to realize when he wrote his gushing
dedication to Charles I. The heart was now appreciated as a regulator
of the body, and became therefore more powerful than ever as a symbol of moral self-regulation. We speak from the heart when we mean
what we say. We keep secrets in our heart. Even though we know the
brain is the centre of perception and cognition, the heart is still where
we wish to *feel* things. In the West, the heart has for long been the
organ most closely associated with the emotions, though in the East
it has often had more to do with the intellect and intuition. Once,
Western beliefs called on the heart to perform these duties, too. 'For
as he thinketh in his heart, so is he,' warns Proverbs 23:7. The prayer
known as the Sarum Primer after the 1514 book in which it is found
runs:

> God be in my head,
> And in my understanding;
> God be in my eyes
> And in my looking;
> God be in my mouth.
> And in my speaking;
> God be in my heart
> And in my thinking;
> God be at my end
> And at my departing.

At this date, the heart is identified with thinking, while the head, or
brain, is concerned with understanding. Ironically, Harvey's discovery a little over 100 years later that the heart was a pump – a central
pump, regally important in the body, but just a pump for all that –
was one of the first breakthroughs to begin to persuade people that
the brain was in fact more important, marking what the cultural

historian Fay Bound Alberti calls the 'scientific transition from a cardio-centric to a cranio-centric body'.

In 1997, a Canadian cardiologist, Andrew Armour, published a paper making the revolutionary claim that the heart actually has a 'little brain' of its own. Neuronal circuits observed on the heart may be capable of 'local information processing', Armour suggested. The heart is here reframed as analogous not to a pump or any mechanical contrivance, but more fashionably to a computer system: the brain is our mainframe while the heart and perhaps other organs too are served by local processors. Dismissed in some quarters as pseudo-science, Armour's findings were seized upon by churches and theosophists as providing scientific evidence for the biblical thinking heart.

One way or another, the heart retains its place in our hearts, as it were. Metaphors to do with the heart seem very real. To die of a broken heart is surely one of the most awful ways to die, never mind that this squashy, elastic organ cannot break in a physical sense. It can weaken, become atrophied and diseased, but it is never the brittle object implied by the cliché of a heart with a lightning bolt cracking it in two. The emblematic status of the heart is assisted by its compactness and portability. Especially in the case of saints and martyrs, the heart was often buried separately from the rest of the body. This practice stemmed in part from necessity – guts and eviscerated organs were buried first in order to lessen the stink of a rotting corpse in church. But it was also symbolic. The heart, as Young tells us, can also be 'pickled, sent, given, kept, eaten, or worn round the neck'. A heart could even be repatriated from foreign wars when plague laws prevented the return of the body.

Given its symbolic importance, it is perhaps surprising that we are happy to remain largely ignorant about the real appearance of the heart. The beating, visceral thing itself plays so invisible a part in our lives that we do not even know its shape. This is true of the human heart and animal heart alike, for the latter has been marginalized in the kitchen, not central at all, but classed with offal. At the same time, the heart has become ever more standardized as a symbol. Drawings of the seventeenth century show the heart shaded as a three-dimensional object, not always delineated with anatomical accuracy perhaps, but nonetheless at least displaying some of the irregular morphology of the real organ. But during the eighteenth and nineteenth centuries, on playing cards, in woodcuts and embroidery, and finally on commercial Valentine cards, the heart became far more familiar as a flattened and symmetric figure.

How did the heart arrive at this stylized, and most unrealistic, two-dimensional device – a red, twin-lobed, inverted triangle? Theories are many and ancient. In Egyptian hieroglyphics, a vase stood for the heart. Is our heart icon the outline of a vase? The curlicued design of a lyre offers a Greek explanation. Or it may simply be a development from that inverted triangle used to represent the female sex, a symbolism celebrated by the fashion designer Mary Quant, who got her husband to clip her pubic hair in this shape. In fact, the design we interpret today as a symbol of the heart had its beginnings as the depiction of an ivy leaf or a bunch of grapes. The symbol on the suit of cards that we call 'hearts' was originally such a leaf.

Hearts in medieval art and literature were often described as pear- or peach-shaped. Giotto's fresco of Charity in the Scrovegni Chapel in Padua has her offering a teardrop-shaped heart from a bowl of fruit. But at some point the flattened ivy leaf motif seems to have taken over as the preferred shape for the human heart. The first heart with a cleft may be that depicted in Francesco da Barberino's book of emblems, *I Documenti d'Amore*, dating from around 1310, while the first stylized heart in an illustrated anatomy dates from 1345. In churches, worship devoted to the Sacred Heart of Jesus gradually supplanted the Franciscan devotion of the five wounds of Christ. Later, the Sacred Heart alone became the symbol of the Roman Catholic backlash against Protestantism. This lurid symbol was not without its problems, however. At the end of the nineteenth century, for example, Catholic missionaries in Rwanda found themselves accused by their would-be converts of cannibalism because of the graphic nature of their crusading logo.

The simplified heart shape was cut into furniture by the Amish and the carpenters of the English Arts and Crafts movement. Today, it features in the branding of many products, promising, confusingly, either that they are good for you or that they are naughty-but-nice. There is even a key option for a heart symbol on my Apple computer, which has served me no purpose until now: ♥.

The New York designer Milton Glaser was the first to put the ♥ in a sentence: I ♥ NY. This long-lived slogan – it dates from 1976 – has succeeded far beyond its creator's anticipations. It sends an unmistakable warm embrace, disarming the city visitor who might otherwise tremble before the urban chaos. I ♥ NY is ingenious above all because it has a truth at its core to do with our love of place and how that in turn creates community. There is a more calculating cleverness about it, too. The logo is easily copied. There are knock-offs of the I ♥ NY symbol everywhere in the city, and this is not accidental. Whereas great effort goes into ensuring that a corporation's logo is only reproduced by the right people in the right way, Glaser's logo has no copyright protection. The idea was that anybody and everybody in New York could use it. It was an unpredictable strategy, but more than thirty years on, it has yielded huge dividends. True, it is not replicated with precision on every occasion. The heart shape may not swell in quite the way of the original; the typeface will more than likely not be the one (American Typewriter) that Glaser chose. But, in its way, the design is doing its job all the better because of this, showing as well as everything else that New Yorkers are nobody's conformists. And there is undreamed-of multicultural diffusion far beyond the five boroughs. Clumsy homage is paid by other states: 'Virginia♥ is for lovers', for example, or 'I L♥Vermont', both official bumper stickers. J'♥ Quebec, Me ♥ Antigua and I ♥ Allah are found further afield. All these variants subliminally recall New York too, effortlessly augmenting the message of many cultures rubbing along together that is such an intrinsic fact of New York life, even as they announce their own passions.

★

The kidney is quite as shapely as the heart. Any self-respecting St Valentine's Day confection must be heart-shaped lest its amatory purpose be overlooked. But today we also find kidney-shaped cakes being made to celebrate successful transplant operations. In the manner of party cakes, these are often gruesomely realistic, sometimes with the ureter and major blood vessels sculpted in colour-coded icing as if copied from an anatomy textbook. The implications of this new custom seem not to have been worked through. The giving of heart-shaped gifts is clearly meant to represent the giving of one's own heart. A kidney-shaped bakery item starts off well enough as a kind of 'rebirthday' cake. In eating it, the recipient perhaps re-enacts the incorporation of the donated organ. But the symbolic consumption of the donated kidney by any other celebrant seems a touch macabre.

Most organs have, like the heart, a shape that is characteristic, but still sufficiently irregular as to elude easy description. In other words, one heart is shaped pretty much like the next, but not enough like any familiar object that it may be used as a visual index. The kidney goes a step further, having a shape that is so characteristic of itself only that it has given its name to a miscellany of other natural and manmade objects, from kidney beans to the garden swimming pools advertised as 'kidney-shaped', presumably designed that way to look more natural than the obvious rectangle.

The leaves of plants, too, are sometimes kidney-shaped, or reniform to use the technical term. There is a single explanation for the occurrence of this unusual shape in so many natural organisms (if not in swimming pools). We have seen how the stylized heart may have developed (and perhaps the diamond, club and spade, too) from diagrammatic representations of different leaves. D'Arcy Thompson, in his masterly work *On Growth and Form*, shows how all these shapes originate from small alterations in the radial and tangential vectors of leaf growth (that is to say, the rate at which growth thrusts upward from the stem and the rate at which it spreads aside). A high thrust and a low spread results in a lanceolate leaf, or 'diamond', whereas a heart shape arises when the spreading force is greater relative to the thrusting force so that part of the leaf fans out wider around the stemming point. Further restriction on the nominal upward growth leads to the squashed but otherwise symmetrical kidney shape seen in the leaves of plants such as pennywort, many beans and our own kidneys.

Various hard-to-describe shapes come to the fore in Vladimir Nabokov's novel *Bend Sinister*. Recurrent visual motifs – puddles oblong and 'spatulate', the water-filled outline of a footprint, an ink stain in the shape of a lake – seem to hint at something of vital importance that has been forgotten by the recently bereaved central character, Adam Krug, who is engaged in a struggle against the totalitarian regime run by his former schoolmate. The tale also abounds in images of human organs – an inflated football has 'its red liver tightly tucked in'; there is 'a black colon' of ink on somebody's collar; a person's rump is like 'an inverted heart'. The shapes and colours, and the memories which they seem to represent, enable the reader to share a little of the synaesthetic condition to which Nabokov was subject. These symbolic strands eventually converge when Krug's tormentor spills a glass of milk, forming a kidney-shaped puddle, providing an unnecessary reminder that Krug's wife died following an operation on her kidney.

There are many mysteries that remain to be uncovered concerning the curious forms into which the body and its organs grow. Not the least of these is the matter of why we possess two kidneys. Nature's general rule is to give us just as much of everything as we need, no more and no less. Two horizontally set eyes give us binocular vision by which we are able to judge distance. The spacing of our two ears likewise helps us to determine where a sound is coming from. The UK National Kidney Federation, however, says it's not known why we have two kidneys. It may be a knock-on effect of the general anatomical doubling that produces two legs very early in the development of the embryo. This would also explain why we unnecessarily possess two testes or two ovaries. Or it may be the legacy of some necessity far back in our evolutionary past. Most animals have two kidneys like us, but some have more, and even the human embryo actually develops three pairs of kidneys about a month after conception, with only the last of the three becoming functional organs.

In the end, neither the shape nor the number of kidneys matters as much as their function. One in 400 of us in fact possesses a single kidney formed by the fusion across a central isthmus of two normally placed kidneys. Such 'horseshoe' kidneys often work perfectly well without producing any symptoms or evidence of their presence. It is typical of the kind of abnormality that can pass utterly without notice because it is internal, and yet which if found on the surface of the body can so easily cause people to recoil.

Its redundancy has made the kidney the trailblazing organ in human tissue transplantation. The kidney that remains in the body of a living donor soon grows by some 80 per cent, practically restoring full renal function. Surgeons at Harvard Medical School carried out the world's first successful kidney transplant operation in 1954, using identical twins as donor and recipient in order to reduce the risk that the organ would be rejected. The recipient lived for another eight years; the donor only died in 2010, at the age of seventy-nine. In the United Kingdom, 2,732 people received a new kidney in the year to 2011, with just over 1,000 of these being transplants from living donors, but there are nearly 7,000 people on the waiting list. In the United States, around 15,000 operations are performed annually, but the waiting list is approaching 100,000 and rising fast. It is estimated that by 2015 this number of patients *each year* will be experiencing renal failure, for whom a kidney transplant may be their only hope.

Broadening the range of donors is fraught with both medical and ethical difficulties. For example, potential donors unrelated to the prospective recipient have been assessed in the past, but found to be borderline 'psychopathological'. 'Emotionally related donors' are thought to be more reliable. Another controversial proposal is to grant clemency to death-row prisoners in exchange for a kidney. This almost Swiftian idea seems tempting when one remembers that there are more than 3,000 inmates facing the death sentence in the United States. However, since this number has remained virtually static since 1996, it seems more of a political gesture than a practical solution.

The idea of transplantation follows readily enough if we believe that parts are discrete and separable from the body that contains them. Greek surgeons carried out experiments transplanting human bones as early as 400 BCE. Reasons for failure were medical – there was no understanding of rejection and the immune system. But there were also powerful moral reasons for hesitation – such as the forcible means by which the body parts were then obtained, and the obvious infringement of the first injunction of the Hippocratic oath to do no harm.

The success of the first kidney transplants in the mid twentieth century was quickly eclipsed by the more glamorous and symbolic transplant of the heart. Being unique, a heart could not be provided by a perfectly matched donor, such as a twin, as with a kidney. Instead, much greater pre- and post-operational care was required in order to ensure a functional result, as well as great skill from the surgeon. Christiaan Barnard, the Cape Town surgeon who became a household name when he performed the first successful operations, practised first on dog hearts, performing more than fifty transplants. (He also grafted a second head on to a dog, simply, it seems, because he could.) Barnard's first human heart recipient survived for eighteen days; the second for eighteen months. After these early successes, however, the image of heart transplantation suffered setbacks when others began to perform the operation with much lower rates of survival, and when some of Barnard's own patients quite coincidentally began to exhibit psychotic behaviour after recovering from surgery.

Today, though, transplants are a standard if extreme option in the surgeon's repertoire. Transplantation is broadly accepted not least for pragmatic reasons because of soaring demand for replacement organs. But it remains, in the words of Columbia University anthropologist Lesley Sharp, 'simultaneously wondrous and strange'. It is a medical procedure, to be sure, but no amount of mechanistic jargon – the heart characterized as just a pump, the liver and kidneys mere filters – can disguise the fact that it is also a personal act, a gesture from one person to another that seems as if it ought at least to obey the usual social rules of giving. As Sharp explains: 'donated cadaveric organs simultaneously emerge as interchangeable parts, as precious gifts, and as harboring the trans-migrated souls of the dead.'

Surgeons and neurologists refute the notion that aspects of personality can be transferred from person to person during transplant operations. But nothing can prevent recipients from imagining things about the donor of their new organ, especially when that organ is the heart. Patients who express the sense that another person dwells within them – only a very few, medical agencies insist – are labelled as victims of 'Frankenstein syndrome'. Fay Bound Alberti gives the example of Claire Sylvia, a heart recipient, who had been a healthy-eating dancer before her operation and inexplicably became a lover of chicken nuggets afterwards. More natural is the guilt that a recipient may feel at having received a replacement organ and given nothing in exchange. Michelle Kline, for one, felt so guilty about receiving her brother's kidney that she was unable to talk to him at all until she had shown herself worthy by becoming Miss Pennsylvania and a finalist in the Miss America beauty pageant. When her brother saw her crowned, he remarked: 'We looked good up there on stage.'

For their part, a deceased donor's kin may feel that the donor's identity lives on in the 'new' body. Donor anonymity rules mean that direct connection is not usually made between the donor's kin and the recipient, but occasional breaches have occurred. Ralph Needham received a double lung transplant from a donor who had died following a severe head trauma. He commented of the donor's wife: 'Her husband gave me two good lungs. She thinks that her husband lives on in me, but I feel uncomfortable about that – I feel they are *my* lungs now.'

The social understanding of an organ as a gift sits uncomfortably with the way that modern medicine actually operates. Though organs are usually transacted within state-run health authorities or non-profit organizations, it's not long before the language of money makes its appearance. Putting a price on human organs is frowned upon to say the least, and trade in them is widely banned, and yet we store them in banks, for example. In fact, a single cadaver can yield

150 usable parts, 'worth' more than $230,000 in all. Though organ donation depends on selfless givers who gain no monetary reward, transplantation is said to be 'among the most profitable medical specialties' in America.

I raise some of these ethical puzzles with James Neuberger, the associate medical director of the UK National Health Service Blood and Transplant authority and a liver transplant surgeon himself. He begins by noting the sharp disparity in attitudes from country to country. 'Where death is more freely discussed and accepted, donation is more accepted, for example in Catholic countries. But in Southeast Asia donation after death is very rare. Whether it's religion or culture, I'm not sure.'

On some aspects of donor psychology, he takes the medical materialist view I expect. 'The concern is for the body, and perceptions of what happens to your body and your organs after death, but when you're dead you're dead, so far as I'm concerned. People don't see what a body looks like after six months – there isn't much of it left.' But then he surprises me with this: 'My personal view is that what makes humans different from the animals is not the body but the spirit.' He is scathing about resistance to donating one's organs based on the idea that one would not then be going to God intact. 'I've never heard that when people have had their tonsils out.' But he immediately tempers the thought, adding that he knows of cases where amputees have wished to be buried along with their preserved cut-off limb. 'The first thing is to know what people really feel, and why they have concerns.'

Neuberger is hopeful that transplantation as we presently understand it may turn out to be a short-lived episode in medical history. At a technology conference in March 2011, Anthony Atala, director of the Wake Forest Institute for Regenerative Medicine in Winston-Salem, North Carolina, described how three-dimensional printing machines of the kind beginning to be used to fabricate customized items in plastic could be adapted to 'print' human tissue. In this case, a patient's wound is optically scanned, and the digital information thereby gathered used to determine the size and shape of the tissue required to occupy the void. This shape is then fabricated by depositing healthy cultured cells layer by layer in a suitable matrix where they can fuse together to form a functional organ. Atala printed out a specimen kidney for the benefit of the conference audience. 'It's just like baking a cake,' he told them.

Blood

From the experimental data that he amassed revealing the awesome power and capacity of the heart, William Harvey concluded with irresistible logic that the blood that pumps through it cannot possibly be generated afresh at the necessary rate, and that therefore it must be carried in a circuit repeatedly around the body. Chapter 14 of his *De Motu Cordis* draws his thinking to a crisp conclusion. It reads in its entirety:

> And now I may be allowed to give in brief my view of the circulation of the blood, and to propose for it general adoption.
>
> Since all things, both argument and ocular demonstration, show that the blood passes through the lungs, and heart by force of the ventricles, and is sent for distribution to all parts of the body, where it makes its way into the veins and porosities of the flesh, and then flows by the veins from the circumference on every side to the centre, from the lesser to the greater veins, and is by them finally discharged into the vena cava and right auricle of the heart, and this in such a quantity or in such a flux and reflux thither by the arteries, hither by the veins, as cannot possibly be supplied by the ingesta, and is much greater than can be required for mere purposes of nutrition; it is absolutely necessary to conclude that the blood in the animal body is impelled in a circle, and is in a state of ceaseless motion; that this is the act or function which the heart performs by means of its pulse; and that it is the sole and only end of the motion and contraction of the heart.

It is exemplary scientific reporting, plainly and fully descriptive, and utterly lacking in the kind of baroque literary flourishes that characterize so much seventeenth-century writing. The circularity especially pleased Harvey, leading him to draw an analogy with the water cycle as described by Aristotle. Before long, the healthy circulation of blood discovered by Harvey would inspire metaphors of other healthy circulations, such as that of trade within the nascent British Empire.

The circulation of the blood began to explain formerly puzzling phenomena such as how an infection in one part of the body could quickly spread to other parts. But traditional views of the blood itself – the red liquid that runs from our wounds and for which our bodies are apparently the container – hardly needed to change at first. The fact that the blood was circulated rather than generated gave no cause to modify established medical treatments such as bloodletting (in which a vein is cut open to release a quantity of blood) and cupping (in which a heated vessel is placed on the skin in order to draw blood to an affected area); indeed, in Harvey's view, the circulation of the blood explained their supposed efficacy for the first time. Harvey's discovery marked a radical shift from the Galenic view in which blood was manufactured in the liver, given ruddy life in the heart, and then sent out to all parts of the body, never to return, like the light of the sun. But this revolution in one of the four Hippocratic humours – phlegm, black bile and yellow bile being the others – did little to upset the balance of this system of medicine, which continued to guide physicians for another couple of centuries after Harvey. Other ancient beliefs to do with the blood – the horror and fear of it, and rituals and taboos surrounding its appearance – all continued intact.

In Judaism, all blood is regarded as the source of life. Animal flesh only is to be eaten; the blood is to be poured away on the ground or poured sacrificially on the altar of the Lord, according to Deuteronomy. Human blood is unclean. The privileging of the blood in this way stems in the view of some anthropologists from a folk memory of human sacrifice, but it is surely also evidence of a primitive awareness that the blood may be infected with disease.

Although Christianity arose out of Judaism, its attitude towards blood is sharply different. Because the Christian God is revealed in the bloody sacrifice of Jesus, blood is a central part of the ritual. Until the Fourth Lateran Council of 1215, the Christian ceremony involving bread and wine was merely symbolic of the Last Supper. The

Council decreed that the bread and wine was to be regarded as the actual body and blood of Christ, and in doing so invented a ritual, the Eucharist, that could be replicated in every church in Christendom, in which the faithful could engage in a physical communion with Christ. Thus is blood seen, meditated upon, and even drunk. By the miracle of transubstantiation, believers can share in the body of Christ without disgust, neatly sidestepping any suggestion of cannibalism. That far older ritual is inevitably what springs to an anthropologist's mind, however, and the Christian altar will always carry a faint echo of the pagan sacrificial table.

Blood is unclean or polluted as soon as it leaves the body. It shares this property with other bodily emissions, such as urine, faeces and phlegm. But it does not *normally* leave the body like these other substances, and so its appearance in the outside world is always remarkable. Often, of course, it is an ill omen. John Keats, the one-time trainee surgeon, recognized his own impending death from tuberculosis at the age of twenty-five when he saw on his pillow 'arterial blood. I cannot be deceived. That drop of blood is my death warrant.' Destined to die from the same disease a century later, Kafka interpreted his blood rather differently when 'in the swimming bath I spat out something red. That was strange and interesting, wasn't it?' A normal stool or a gobbet of phlegm is neither strange nor interesting. But blood demands notice.

Men found menstrual blood especially disturbing. A penance of some weeks of fasting was traditionally demanded of women who entered a church while they were menstruating. The 'churching' of women was a ritual of forty days' 'decontamination' of a new mother after giving birth, during which she was required to withdraw from the church and from society, a custom followed in some places well into the twentieth century. The sexual inequality is set from birth: according to Leviticus, a baby girl is double the trouble, rendering the mother fourteen days unclean compared with seven for the birth of a boy. Menstrual blood was feared as a reminder of the uterus, the organ of female fertility that might so easily form an alternative basis for worship to the elaborate system erected by the male priesthood. Menstrual blood is not a universal taboo, as the anthropologist Mary

Douglas demonstrates with reference to the Walbiri people of central Australia, whose women are subject to brutal physical control by their husbands, apparently obviating the need for more nuanced rules of sex pollution. But it was and remains widespread (think of tampon advertisements that puzzlingly use blue ink to demonstrate their efficacy). In general, the appearance of blood is a sign of weakness and ineptitude in man, as when he is wounded in battle or, more likely these days, cuts himself shaving. But in women it is a reminder of life-giving strength, and in male-dominated societies this leads to social division, expressed for example in the slander that contact with a menstruating woman has the power to tarnish mirrors, sour wine, stifle infants in their cradles and fatally weaken a man in all sorts of unpleasant ways.

I find a few of these observations in *A History of Women's Bodies*, written by a man, Edward Shorter, 'the somewhat lurid title' chosen, he announces epiphanically in his preface, 'to make the point that women's bodies have a history of their own'. My library copy of this 1982 work has been liberally annotated with expressions of incredulity by recent generations of women students, not so much at the tales themselves but at Shorter's constant problematizing and medicalizing of his subject, women's bodies, which seems in its way to perpetuate ancient patriarchal prejudice. Nearly half the book is given over to the subject of childbirth, for example, and one chapter is entitled 'Did women enjoy sex before 1900?'

Before we knew about genes, blood was understood also as the medium of our heredity. Blood is family. 'Am I not consanguineous? am I not of her blood?' demands Sir Toby Belch in *Twelfth Night* in reference to his niece Olivia. Blood is also tribe. 'For blood of ours, shed blood of Montague,' exhorts Lady Capulet in *Romeo and Juliet*. And blood is race. Racial purity is often gauged in terms of blood, as was the case with the notorious 'one-drop rule' adopted as law in many southern American states in the early twentieth century. Under the rule, any person with the slightest African heritance ('one drop' of blood) was legally defined as black (in more liberal states, one-eighth or one-quarter African ancestry was the definition). Enforcement was impossible, of course, and in practice court cases drew on evidence of recent ancestry. Genetic tests today suggest that more than a quarter of 'white' Americans would fail the one-drop rule.

I find that many of these old beliefs seem to persist when, for the first time, I enrol to give blood. First, I must complete an online questionnaire. By doing this, I consent to allow 'medical, religious or other sensitive personal information submitted by me to be used by the National Blood Service'. The form asks many of the expected questions about my general health and likely exposure to infection. There is also a section about 'lifestyle', demanding to know about my likelihood of exposure to HIV and hepatitis viruses, and whether I have had acupuncture, piercings or tattoos, as well as probing gently at my sexual tastes. There are several questions which it is impossible to answer with complete certitude, such as have I ever 'had sex with anyone who has ever injected drugs', or have I 'had sex with anyone who may ever have had sex in parts of the world where AIDS / HIV is very common (this includes most countries in Africa)?' Neither can I be totally sure that I have not, in the last four weeks, 'been in contact with anyone with an infectious disease'.

Any new disease immediately prompts the fear that it is carried in the blood. Scientists were at first highly reluctant to believe that AIDS was carried in blood, because of the awful implications for infectivity. Conversely, once a particular infection *is* associated with the blood, it may prove very hard to revise the general opinion. In Canada and other places, declared gay and bisexual men have been debarred from giving blood. However, more effective methods of screening donated blood for HIV and hepatitis, and the reduced likelihood that such men are carrying these viruses, owing to better education, have now led the Canadian authorities to contemplate relaxing the ban. But first, to see whether this would be a wise move, further research was required, for which half a million dollars' funding was offered. Most untypically, no scientists came forward to take on the work.

Odder still are the questions on my form about 'Travel outside the

UK'. These seem to assume that the nation's borders should be impervious barriers to disease and impure blood. They put me in mind of John of Gaunt's speech in *Richard II*: 'This fortress built by Nature for herself / Against infection and the hand of war'. I am asked whether I have been abroad within the last twelve months, and made to feel it is somehow improper that of course I have. The questionnaire also demands to know if I have 'ever lived or stayed outside the UK for a continuous period of 6 months or more'. My national loyalty is again found wanting. I tick the yes boxes, and the online questionnaire promptly shuts down, thanking me for my trouble with this confusing consolation: 'You may still be able to give blood.' Out of curiosity, I re-enter the site and lie my way through to the end. This time, it rewards me with the message: 'It seems you are able to give blood', which I interpret as their way of saying: 'We think we can accept your blood.'

I wonder what will happen to my blood if I am allowed to donate. Will it be mixed with the blood of people of other ethnicities, of foreign parentage, lovers of exotic holidays? Is the tendency of health policy towards a global blood bank that recognizes our common humanity (while distinguishing between blood groups to ensure antibody compatibility)? Or is the countervailing movement stronger, and which I have heard is growing, especially in the United States, for people to build up banks of their own blood for their exclusive use?

On the day of my appointment, I make my way to my local town hall. There, half a dozen couches are laid out, with people in blue uniforms bustling around them. I sign in, and am encouraged to help myself to a large glass of water or sugar-free juice. I have been relaxed about the idea of becoming a blood donor up until this point, but now I find I have butterflies in my stomach, and my left arm is tensing in anticipation of the needle. The majority of the donors on this day are women. Ages seem to span the full range permitted for donors, seventeen to sixty-six years old. I sit down to wait, idly flipping through leaflets that seek to reassure me as to what I am about to undergo. One has a photograph of a mournful spaniel on the cover. Puzzled as to how it can possibly be relevant, I open it and read that the first blood transfusion, noted by Samuel Pepys in his diary, took place in 1666, when, according to the minutes of the Royal Society, a spaniel received blood from a 'little mastiff'. The copy continues with more honesty than judgement: 'The spaniel survived (although the mastiff was less fortunate), and scientists were encouraged to move on to human subjects.' I'm just wondering quite why everybody felt so chipper given the demise of the donor dog when my name is called.

A nurse first runs through my questionnaire responses. We discuss the answers I have left blank. I explain that I have been abroad – to Italy and the Netherlands. I'm in the clear: had I been to the northeast of Italy or one or two other places, I might have been ruled out because of the risk that I had picked up West Nile virus. I had hesitated, too, over a question about hospital operations. Did out-patient treatment to have my wisdom teeth removed count? What about the time they set my broken leg? This requires the nurse to consult with a colleague. Finally, I am judged acceptable, and passed to a second nurse, who checks my blood density by placing a drop of it in a solution of copper sulphate. This will confirm whether I have at least the average level of iron in my blood, which is the threshold for donation. The drop hovers, then sinks. I have passed.

I am sent over to one of the couches, where a third nurse inserts the needle in my right (not left!) arm. The scratch is almost completely painless – certainly more skilfully done than when I last had a blood sample taken by the nurse at my GP surgery. Then she starts the machine that over the next ten minutes or so will withdraw 470 millilitres of my blood. I feel a sensation of warmth where the tube carrying the blood is taped to my lower arm – my own departing body heat. Soon the plastic sac is plump with dark red liquid. The quantity is not quite the pint of legend – 'that's very nearly an armful', as the comedian Tony Hancock had it. As I lie there gazing at the roof of the town hall, I ask the nurse what effect she reckons this classic sketch has had on blood donation in Britain. She gives a mirthless laugh and says nothing.

After it is over, I am invited to rest for a few moments and drink more fluids. Regular donors are sitting around comparing notes as to when they first gave blood and what inducements or motivations they had for starting. The local vicar is among them, tucking in to the biscuits. I wonder how much my blood is really worth. Collection is an intensive effort. There are more than a dozen staff at this station. Their target for the session is 115 donors, which will give them something over fifty litres of blood. Is a glass of juice and a three-pack of bourbons fair recompense for my contribution? I ask what happens next to my blood – or the blood that I am still calling mine. I am told it is taken to the National Health Service Blood Transfusion centre in north London to be tested and stored ready for use. Later, I find there is a price put on my blood. The internal market in the NHS means that it is 'sold' by NHSBT to hospitals – my 470 millilitres has a price tag of around £125. That's a lot of biscuits.

I leave the town hall. Does the sunshine seem brighter? The air sharper? I am not sure. Am I light-headed, as I was warned I might be, or is it just the natural adjustment from having been indoors on a glorious spring day? A couple of weeks later, I am impressed when I receive a personal telephone call of thanks. There follows a standard letter confirming my blood group and telling me I have 'done something truly amazing'. With it comes what I assume is a kind of loyalty card, which I am invited to carry with me. It is red and says I have made '1–4 donations'; the highest grade is for those who have made more than 100 donations.

Many academic studies have examined this unusual transaction – a gift system in which donor and recipient are mutually anonymous, in which not all may give, in which no reciprocal gift is exchanged. When asked, donors tend to claim humanitarian motives and altruism as reasons for giving blood, but an underlying selfish satisfaction has also been shown to be a strong factor. I can believe this when I return for my second appointment four months after the first. On this occasion, my blood iron comes in just too low, and I am sent home. The feeling of rejection is surprisingly acute.

Giving blood is a humanitarian act, and one 'taken in the face of obvious physical costs', according to one study. It is more obviously a sacrifice than giving money to charity or helping an old lady across the road. Yet giving blood, it is suggested, can become part of oneself, part of what defines a person. The authors draw a comparison with church-going. In one survey asking first-time donors why they chose to give blood, a maintenance fitter is to be found quoting John Donne: 'No man is an island.'

It is important to know more about donors' motives in order to promote donation. The standard letter I received from NHS Blood and Transplant tells me that only 5 per cent of people who could donate do so regularly. Repeat donors are worth far more to the health service, although research has focused almost entirely on first-time donors. Money, it seems, is not the answer. Making the exchange a commercial one – giving a fee instead of biscuits – is generally thought to run counter to the high-minded, communitarian motives identified in surveys. It would also be likely to skew the donor profile towards those who need the money, and this in turn would prompt (not always rational) questions about the quality of blood likely to be obtained. In the United States during the 1950s and 1960s, 'Cash paid for blood' signs appeared in slum neighbourhoods, and donors were encouraged to give blood in exchange for family credit. Yet the number of donors rose only modestly at this time, while in Britain following the founding of the NHS the rate of donation increased almost fourfold.

Donor enrolment has continued to rise but at a slower pace since, raising the question of how to increase supply in the face of rising demand for blood (although in fact it is also known that the *perception* that there is this ever-present need is a major factor causing first-time donors to come back, and this is exploited by donation agencies). There is talk of finding new ways to increase levels of giving. But there may be limits on what is achievable that go back to our deepest fears about our life-blood. In the 1960s, the notorious American euthanasia activist Jack Kevorkian known as 'Dr Death' – and also the author of a musical suite entitled *A Very Still Life* and sometime painter in his own blood – proposed that blood be harvested from fresh human corpses. His early experiments confirmed that cadaveric blood could be used in transfusions, but his work was rejected by hospital colleagues. He suggested in the journal *Military Medicine* in 1964 that the technique might be of use on the battlefield, but failed to interest the Pentagon. The idea might be regarded in principle as no more objectionable than harvesting organs from certificated donors. Blood is just another tissue, after all – one of the connective tissues, so-called because it runs throughout the body rather than being associated exclusively with localized organs. But it seems the cultural barriers are greater than the medical ones.

Perhaps my selfless civic gesture is a relic of the old custom among country people to submit to a bloodletting each spring and autumn equinox. This tradition owes much to Galen, whose thinking shaped both Islamic and Western medicine for many centuries following his death around 200 CE. Strengthened by medieval attempts to link aspects of human health with the astrological calendar, it became a seasonal ritual which did not entirely die out until well into the nineteenth century. It was usual to let, or drain off, about as much blood as in the sac that I filled for the NHS. I had seen the gory tools used for this in medical museums – the simple lancet, and a scarifying device like a miniature version of a spiked roller for aerating a lawn that could inflict many small, shallow wounds across an area of skin. Bloodletting persisted for centuries, not least because it often worked. It was a practical remedy for people suffering from high blood pressure, heavy menstrual bleeding, haemorrhoids and various inflammations and fevers. It was doubtless also effective as a placebo – as pills are today – as well as having the salutary spiritual effect of inspiring thoughts of Christ on the Cross.

This is not to say there were not occasions where letting blood was absolutely the wrong thing to do. On 14 December 1799, George Washington awoke with a severe cold in the throat. A servant prepared a balm of molasses, vinegar and butter, which the general was unable to swallow. Instead he demanded that half a pint of blood be taken from his arm, having previously used bloodletting to good effect on his slaves. When his physicians arrived, they continued the treatment, the first on the scene draining forty ounces of blood in two venesections, the second taking another thirty-two ounces. It was, as both his servant and his wife Martha had feared, the worst treatment, and by that evening the first president of the United States was dead. He can have had hardly any blood left in his body at the end.

Blood was first among equals of the four humours, the system that governed medical practice for more than two millennia, from before Hippocrates to well after the rise of anatomy and modern medicine with Vesalius and Harvey. The four humours were invisible themselves, but their signs were seen in the blood, phlegm, yellow bile and black bile. Blood carried all the humours through the body as well as the thinner fluid known as the spirit. Phlegm included a range of protein-rich secretions such as tears and sweat as well as nasal mucus. Yellow bile was seen in the pus that appears around a healing wound and in stomach juices. Black bile was seen in congealed blood or an abnormally dark stool. The four humours represented balance in the human microcosm as the four elements (earth, air, fire and water) and the four seasons did in the macrocosm. Illness was regarded as an imbalance in one or more of the humours. Cholerics had an excess of yellow bile associated with the same hot and dry temperaments as the element fire. Phlegmatics were their opposite, cold and damp, like water. Melancholics combined dry and cool, like the earth, while sanguine types were warm and damp, like blood and the air. (It is perhaps an indication of the logical power of this neatly interlocking system that its obvious basis in a Mediterranean climate could be so easily overlooked by physicians in northern Europe, where the air is cold and the earth damp.)

Doctors had no way to supplement a deficiency in any of the humours, so in practice treatment was based on relieving a supposed excess. Bloodletting was the most direct method. An excess of phlegm could be treated with expectorants, yellow bile with emetics, and black bile with laxatives. The whole body became a kind of hydraulic network of channels, overflows and water gates, from which excess fluid must be run off to keep it in good order. The four humours might seem vague and inadequate to us now, but the system that they described was both self-consistent and robust, as its great longevity shows. It was also evidence-based to a remarkable degree. I am left with the conclusion that an equinoctial bloodletting is more likely to do me some good than a modern-day detox, although I think I prefer the rituals of the NHSBT over submission to the scarifier.

The idea of the humours lives on, not only in our continuing belief in the diagnostic value of a blood test, the sound of a cough, the appearance of a stool or of a cut as it heals, but also in our recognition during the past century of the importance of endocrinology. It is now the endocrine system, and the hormones that it releases into our bloodstream, that we understand as governing our metabolism and moods, and we have begun to talk about dopamine and melatonin, endorphins and adrenaline in the same way that the ancients spoke of the unseen humours.

The Ear

There is a breed of physician whose members I imagine, bow-tied in all probability, stalking the art galleries of the world, searching out artistic depictions of their medical speciality. Usually, they have retired from practice, but they still maintain the fascination with the particular part of the body that brought them into medicine in the first place. So former hepatologists consider the accuracy of Classical scenes showing Prometheus chained to a rock with the eagle pecking at his liver (marvelling perhaps at the nightly regeneration of the organ so that Prometheus must go through it all again the next day). Podiatrists triumphantly spot the artists who, for reasons of compositional elegance or sheer inattention, have given their subjects two left feet. (It happens more often than you might think.)

Inspired by their example, let us consider the strange case of the ear in Dutch art. The ear enjoys a peculiar status in the art of portraiture. It's generally thought that the hand is the hardest part to draw. Of the facial features, the eyes, nose and mouth are almost unavoidable in any portrait. The ears, though, are somewhat optional. The ears (or more usually, ear) can always be covered by the brim of a hat or an extravagant ruff collar. The ear, it is implied, is an obligato exercise, something that should only be attempted by the most ambitious draughtsman. Even in Vermeer's *Girl with a Pearl Earring*, only the lobe of the ear from which the famous earring hangs is visible. The gallery-going physicians who have set me off in this particular direction, and to whom I am most grateful, are a plastic surgeon, Wolfgang Pirsig, and a medical historian, Jacques Willemot, who together have edited a volume called *Ear, Nose and Throat in Culture*. They confirm this supposition. 'It is not surprising to find prominent ears in the works of four outstanding painters who drew most of their portraits with considerable fidelity to nature: Hieronymus Bosch, Leonardo da Vinci, Albrecht Dürer and Rembrandt,' they write. These artists were notably great draughtsmen as well as great painters.

For most artists, the ears don't really matter – they are a side issue, quite literally. When I drew the partial dissection of a head, I found the ear unavoidable. I reckoned I had made a fair stab at its outline curve and spent considerable effort on shading the folds, and was pleased with my effort until I noticed that I had located it a good inch too far down on the head. The old masters don't make such elementary blunders. But they do treat the ear as an unusually malleable and mobile appendage. Indeed, artists often develop a 'signature ear' by which work may be identified as theirs – a habit that arises from, but scarcely does justice to, the fact that ear shapes vary more greatly between individuals than many other body features.

The star of Jan van Eyck's 1436 *Madonna with Canon van der Paele*, which hangs in the Groeninge Museum in Bruges, is definitely not the centrally enthroned Madonna or the baby Jesus on her lap. It is the other eponymous subject. The canon is being presented to the Virgin in the setting of the church of St Donatian in Bruges. The Madonna is featureless in appearance and bland in expression, but the canon is full of character and wear. He kneels grumpily on the right of the painting. He has just removed his spectacles (then still a novelty) to reveal a face riven with creases and scars. His set jaw juts through a sea of wobbling jowls. An implacable gaze issues from watery eyes weighed down with heavy bags. He is humbly dressed compared with the Madonna and with St George in armour and St Donatian in brocaded robes and mitre who are attending her. With his plain clothes and crumpled visage, he seems almost as if he has been cut out of a modern photograph and inserted to make a profane collage.

The portrayal was once even more grimly realistic. It seems that Canon van der Paele had a large wart or tumour on his left ear. To his credit, he poses for the artist so that we see his left side. And to *his* credit, the artist painted the wart. In this and other works, it is obvious that van Eyck relished the ugly details of human physiognomy. But today, the wart is not there. Nor is there any explanation of its disappearance. It has been airbrushed from history. According to one account, the wart was overpainted during a restoration of the painting in 1934, but the museum does not respond to my requests for confirmation or denial of this story.

Warts are now known to be the result of viral infection, but up until the seventeenth century they were often associated with witchcraft, so there might have been good reason to omit one from a portrait from the start. Undoubtedly many more warts have been left out of great paintings than have been left in. The retention of such a blemish shows a commitment to truthful realism typically representative of the non-hierarchical ideals of the Northern Renaissance. The most famous wart in art is also by an artist of Netherlandish parentage, Peter Lely, who became highly successful in England, managing the trick of working for both Charles I and, after the Restoration, Charles II, and producing a memorable portrait of Oliver Cromwell in between the two. It is not sure that Cromwell asked to be painted 'warts and all', as legend records. Horace Walpole, writing a hundred years after the event, is the source of this quotation. He transcribes an instruction to 'remark all these roughnesses, pimples, warts and everything as you see me, otherwise I will never pay a farthing for it'. Cromwell did have a prominent wart on his lower lip, as both Lely's portrait and Cromwell's death mask attest.

Rembrandt is renowned for the equally unflinching series of self-portraits which he painted throughout his career. In 2003, another of our medics-turned-art critics, Ben Cohen, a retired ear, nose and throat surgeon, noticed that in many of them the artist's ear appears swollen and damaged. In later portraits, an earring hangs from the undamaged lobe below the hardened tissue of the injured part of the ear. Cohen speculates that Rembrandt was the victim of a botched attempt at ear piercing, but nevertheless went back to get the operation done properly later. 'He must have been a very determined young man to risk further damage to the ear,' Cohen writes. As with van Eyck's depiction of Canon van der Paele's deformed ear, the picture provides evidence of artistic honesty. Rembrandt could easily have avoided depicting the damaged ear by painting himself from the other side.

Ears acquire a life of their own in *The Garden of Earthly Delights*. The triptych by Hieronymus Bosch, painted around the turn of the sixteenth century, is well known, but is so crammed with chaotic goings-on that its details tend to escape proper scrutiny. Its three panels

show Adam and Eve in the Garden of Eden on the left, a central scene
of paradise crowded with cavorting naked figures, exotic birds and
oversize fruit, and a dark scene of hell on the right. This is one of the
most fantastical visions in all painting. We see man-eating monsters,
acts of bestiality, severed heads and hands and feet, fire and excrement,
and even a pig in a nun's headdress, illustrating a uniquely ingenious set
of punishments for the seven deadly sins.

Prominently positioned in the picture, just above the central figure
of a man with tree trunks for legs and a broken eggshell for a body
who may well be Bosch himself, is a large knife clamped between a
giant pair of ears. The arrangement of the ears with the knife thrust
forward between them cannot help but suggest male genitals. Lodged
within the folds of the ear that faces us is a miniature black figure who
appears to be lifting another man in. They may be invading demons,
as earrings were originally worn to ward off demons from this orifice,
and this ear is unadorned and therefore unguarded. In his free hand,
the black figure holds a spear that pierces the flesh of the ear. Both ears
are also separately pierced through by a large arrow.

What does this complicated device mean? (Absurdly, you can buy figurines of it, with or without the miniature invaders, from the National Gallery in London, although the painting is actually at the Prado in Madrid.) Bosch's painting goes far beyond the narrative realism developed by northern European artists in an effort to bring biblical stories home to ordinary parishioners, and creates its own nightmare world worthy of Sigmund Freud. Individual details clearly speak to particular sins. Gluttony and lust are severely punished. We see figures vomiting and one unfortunate with his anus on fire as it passes a succession of black birds. Perhaps the ears are severed in order to cut the flow of gossip and eavesdropping that feeds envy and rage. Who, guilty of such a sin and seeing these perforated ears, would not feel their own ears begin to sting?

Bosch's hell is also a clamorous place, so full of musical instruments that some of its denizens are stopping their ears against the din. Physicians note that the giant ears, however, lack auditory canals – they cannot hear anyway. The inner and middle ear contain the mechanisms by which we are able to hear, while the outer ear or auricle acts merely as a sound collector, a point brutally made in Quentin Tarantino's film *Reservoir Dogs*, when the gangster Mr Blonde severs the captive policeman's ear and then speaks into it to see if its owner can hear him. The auricles capture and direct sound into the inner ear – when people with jug ears request surgery to make them less prominent, they are actually likely to hear a little less well as a result. René Magritte (Belgian, not Dutch) graphically suggests how sound might be funnelled into the ear in a small gouache no less surreal than Bosch's image. His *Untitled (Shell in the Form of an Ear)* shows a giant vermiform shell lying on a beach; its spiralling recursions are modelled on the human auricle.

The English colloquialism 'a word in your shell-like', typically used by somebody about to vouchsafe a confidence, acknowledges this similarity. In fact, your ear is more shell-like than you may realize. Our ability to hear sounds of different pitch relies on the cochlea, a hollow bone shaped like a snail shell in the inner ear. It works a bit like a French horn in reverse. There are tiny hair cells all along this tapering tube, tuned according to where they are positioned, like the strings of a piano. They vibrate in response to different frequencies of sound transmitted into the cochlea by the action of the eardrum on the three tiny bones of the middle ear. These vibrations trigger nerve signals to the brain which we interpret as sound. It is amazing that these thousands of hairs can do their job simultaneously so that, in the Haydn symphony I am listening to as I write, I am able to distinguish each instrument that is playing by its individual pitch and timbre. When we lose the ability to hear high-pitched tones with increasing age, it is because some of these hair cells die off. My mild tone deafness, on the other hand, is not explained by physical shortcomings in the ear but a relative underdevelopment of part of my brain, which could probably be remedied with suitable aural exercises if I ever found the time.

Some people believe the outer ear may be significant other than as a sound-gatherer. In the 1950s, a French doctor and acupuncturist named Paul Nogier noted that it resembles a curled human foetus (the lobe of the ear represents the head and the interior fold known as the antihelix the spine of the foetus in this case). The scheme of alternative medicine that he devised based on this resemblance is known as auriculotherapy. The patient's ear is seen as a homunculus or map of the whole body, with stimulation at different points on it being used to treat ailments in corresponding parts of the body. The idea is perhaps not dissimilar to the ancient Greek belief that the ear provides a channel, via the mouth and throat, into the core of the body, and has echoes of the cauterization once used as a treatment for gross body pains such as sciatica, in which some of the flesh of the ear might be burned away using a hot iron. Noting that the ear we can see in Bosch's painting is pierced in two places, Nogier and a colleague took these for acupuncture needle positions and tested the effect on some of their patients. They claim that stimulation at the entry point of their acupuncture needle was effective in suppressing the libido, while stimulation of the exit point heightened sexual interest.

The long hairstyles of the time meant that Antoon van Dyck (who prospered in England during the second quarter of the seventeenth century, going on to be knighted as Sir Anthony) painted many portraits but few ears. One of the exceptions is in an early work, done when the artist was nineteen years old, showing the moment when Christ is captured in the Garden of Gethsemane after his betrayal by Judas. The painting is a chaos of violent action. In the foreground, the apostle Peter is holding Malchus, the servant of the arresting high priest, to the ground and is about to slice off his ear in an effort to prevent the arrest. The ear in question glistens redly as if anticipating the cut. Van Dyck's Peter wields a short knife rather than the sword mentioned in the Bible, which makes the action seem more like a common street crime close to home. Jesus warns Peter to put away his blade (uttering the familiar 'all who take the sword will perish by the sword', in Matthew's gospel). In the version according to Luke, who was a physician, Jesus also heals the ear by a touch of his hand, the only occasion in the whole of Scripture where he heals a fresh wound. This biblical episode has been a stock subject for artists. Most do not waste the opportunity to show Peter in the bloody act, but a few show the aftermath, either with the ear held triumphantly aloft or with the focus on Jesus's remedial gesture. Wolfgang Pirsig has counted fifty-four such paintings, with the artists often choosing freely between left and right ears according to compositional convenience, even though two of the gospels specify that it was Malchus's right ear that was severed. In three careless examples, Christ the healer is actually implanting the wrong ear into the wound on the side of Malchus's head.

Although it seems likely from the way the gospels describe the incident that Peter would have inflicted a mortal wound on Malchus, the taking of an ear has long been a favourite punishment. The War of Jenkins' Ear – now an almost forgotten episode in British history – began with just such a forfeit. In 1731, Spanish coastguards boarded a British merchant ship, the *Rebecca*, making passage past Havana on her way home from Jamaica. The coastguard captain sliced off the left ear of Robert Jenkins, the *Rebecca*'s master, returning it to him to keep as a warning of what would befall other British ships that came his way. Jenkins did indeed tell his story to the king's secretary at Hampton Court when he got home, but the matter lapsed. The actual orifice only became a *cause célèbre* seven years later, when relations with Spain were deteriorating over the matter of slave trading rights in the American territories. Jenkins was called before a committee of the House of Commons in March 1738 and supposedly produced the long-preserved ear in a jar. Jenkins was in fact a reluctant witness, having declined the first summons, and it seems likely that the committee was stacked with Members of Parliament lobbying in favour of war. Detailed records were not made of the committee session, and if Jenkins exhibited anything, it may well not have been his ear but simply a convenient piece of pig meat thrust into his hand as a prop by a lobbyist. Nevertheless, the ear – missing or present – offered a useful symbol of Papist Spanish cruelty to lay before the British public. Britain declared war on Spain the following year.

We can lay the blame for the story on the over-active imagination of the Tory historian Thomas Carlyle, who coined the phrase 'the war of Jenkins' ear' in his monumental history of Frederick II of Prussia, published in 1858. He refers to Jenkins producing 'his Ear wrapt in cotton: – setting all on flame (except the Official persons) at sight of it'. 'Official persons' get the last word, though. The topic of Jenkins' ear appears as one of the frequently asked questions on the Houses of Parliament website, where it draws the cool response that 'it seems highly improbable that he would have kept it for seven years!'

Barbaric ear severings continue to the present day. J. Paul Getty III had his right ear cut off when he was kidnapped by Italian gangsters in 1973. His captors demanded a large ransom from the Getty family, who refused to pay. 'If I pay one penny now, I'll have 14 kidnapped grandchildren,' said Getty's famously mean grandfather. After three months of stalemate, Getty's captors cut off the ear and sent it, along with a lock of hair, to a newspaper, and reduced their demands. The grandfather reportedly now paid up $2.2 million, 'the maximum that his accountants said would be tax-deductible'. Four years later, Getty underwent surgery in Los Angeles to create a prosthetic auricle using a segment of cartilage taken from his rib.

★

The best-known ear in Dutch art is of course the left ear of Vincent van Gogh, a missing part that looms so large it often seems to block our view of the man's pictures. It has become what the critic Robert Hughes disgustedly calls 'the Holy Ear'.

A few days before Christmas 1888, van Gogh quarrelled with his friend Paul Gauguin, whom he had persuaded to come and work with him in Arles. The Dutchman brandished a razor before the men went their separate ways. Later, van Gogh cut off his left ear and gave it to a prostitute named Rachel. 'Keep this object carefully,' he begged her. What she was to make of the gift is unclear, as is what happened to the ear after that. Van Gogh went home to bed, where the police found him the following morning lying almost unconscious on a blood-soaked pillow. This, at least, is the official version. However, an investigation by two German art historians has opened up another possibility. Hans Kaufmann and Rita Wildegans believe that it was Gauguin who inflicted the damage using a sword during the fight, and that the two artists subsequently agreed on the (slightly) more plausible cover story. It is, after all, Gauguin's written accounts that provide most of the evidence for the official story.

What this still doesn't explain is why, if it was van Gogh's left ear that was severed, it is the right ear that appears extravagantly bandaged in his *Self-Portrait with a Bandaged Ear* done a month or two later. Another self-portrait from 1889, produced when the artist had somewhat recovered his mental health, shows him in three-quarters profile from the left side – with ear intact. When newspapers reported Kaufmann's and Wildegans's theory, several of them showed the portrait with the bandaged right ear while blithely referring to the severed left ear in the accompanying article.

The explanation, of course, is that van Gogh painted himself from his image in a mirror. In both paintings, he wears the same greatcoat, which is buttoned at the top. The button is on the left and passes through the buttonhole on the right – the fashion for women's coats. Men's coats normally button the other way round, so this confirms that van Gogh worked from a mirror image. So does a sketch done by his friend Dr Gachet, showing van Gogh on his deathbed with the damaged tissue around the left ear clearly visible.

Using a mirror seems a straightforward enough thing to do. Self-portraitists had been using mirrors since they became generally available, which, perhaps not coincidentally, is when the genre began. Rembrandt, for instance, was suddenly able to paint larger self-portraits when he acquired a larger mirror. But the practice raises a deeper question about identity. Doesn't it matter that we are being shown left for right and right for left – if not to the viewer, who may be none the wiser, then at least to the artist himself? In this very obvious optical sense, the painting does not represent the artist's true self. Jan van Eyck, painting Canon van der Paele, shows us the ugly truth of his damaged ear. Vincent van Gogh, painting himself, shows us a reflection of the truth, but perhaps a deeper truth, too. The common artist's deceit of using a mirror is made to appear less a matter of unthinking procedure and more a matter of deliberate assertiveness by the striking asymmetry of his injury. It was relatively unimportant to van Gogh, as it was to so many self-portraitists before him, that he leave us with a mirror image of his 'true' self. It was more important that he show us his injury: in January 1889, the self-harm *is* the self-portrait.

What are we to conclude about the ear from this little gallery? We have seen the ear as a site of ugliness and imperfection, as a symbol with many meanings, as an object of punishment, as a love token and as a badge of self-loathing mutilation. It is the plasticity of this modest appendage that enables it to perform so many roles. The outer ear is composed entirely of soft tissue and cartilage. There is no supporting bone. This means that it can be deformed and reformed, cut away and replaced. It is an exemplar of arbitrary human tissue.

In this capacity, the flesh of the auricle may stand in for the whole body, either dead or alive. The Mimizuka monument in Kyoto – little known even among Japanese – contains the heaped-up severed ears of Koreans taken as trophies during the Japanese invasion in the 1590s; ears were taken in lieu of heads simply because so many were slaughtered, up to 126,000 according to one source. The removal of an auricle from a living subject leaves only a small wound and cuts no major blood vessels and so is unlikely to lead to death. It too may stand for the whole person, as we have seen with the unfortunate Jenkins.

The function of this fleshy efflorescence in assisting hearing seems almost marginal – it is the inner ear that does the real work. The auricle comes across as a bit of a perk – an erotic bonus perhaps, somewhere useful to hook spectacles, or simply an ornament. It is flesh as a sculptural medium, a notion encouraged by its delightfully baroque curves. The crinkles that complicate the outer curve of the ear arise, incidentally, from an embryonic feature known as the six hillocks of Hiss. Some of these hillocks tell almost forgotten stories. A malformation called Darwin's point, for instance, is the vestigial remnant of a fold that once allowed the auricle to flap closed over the opening of the ear canal, while another of the hillocks was once associated with criminality, and is still sometimes the focus of requests for cosmetic surgery.

These ideas flourish in contemporary art and science, where the ear remains a locus for demonstrating technical prowess. As the art critic Edwina Bartlem observes: 'Strangely enough, artificial ears are powerful signs of tissue culture engineering and biotechnologies more generally.' The iconicity of the ear was only reinforced in 1995 when Charles Vacanti of the University of Massachusetts and Linda Griffith-Cima at the Massachusetts Institute of Technology successfully engineered animal tissue in the shape of a human ear on the back of a live mouse. The growth had no hearing function. It was simply tissue that grew, nourished by the mouse, on a polyester scaffold that could have been made in any shape. So why an ear? One purpose of the experiment was to show that cartilaginous structures could be grown of a kind that in future might be suitable for use in ear transplants. The sculptural delight of the shape may also have played a part. In addition, the instant recognizability of an ear immediately enabled lay people to see the potential of the technology. Perhaps, too, the scientists hoped to generate a little shock value. In any case, the mouse-with-the-ear-on-its-back swiftly became a symbol not so much of the potential to engineer replacement human body parts as of the kind of silly thing scientists can get up to when left to their own devices.

In 2003, an Australian group called Tissue Culture & Art, based at the University of Western Australia, began work with the artist Stelarc on a piece called *Extra Ear ¼ Scale* that seems at once to parody and to extend this work. The idea was to grow a quarter-scale replica of Stelarc's ear from *human* tissue. Oron Catts and Ionat Zurr, the artists behind Tissue Culture & Art, wrote of the original exercise in Massachusetts: 'We were amazed by the confronting sculptural possibilities this technology might offer. The ear itself is a fascinating sculptural form, removed from its original context and placed on the back of a mouse; one could observe the ear in all of its sculptural glory.' Stelarc has since gone on to construct a life-size 'ear' grafted on to his own forearm. The procedure required surgery first to grow extra skin on the arm and then to insert a porous polymer scaffold that would bond with the new tissue in the appropriate shape. Surgeons have performed similar operations as a part of procedures to rebuild damaged ears. Stelarc, however, has gone beyond cosmetic surgery with a disturbing attempt to supplement the body's quota of functioning organs. The finished *Ear on Arm* incorporates a microphone as well as additional electronics to transmit sound and communicate via a Bluetooth connection, allowing people in remote locations to hear what the 'ear' is 'hearing'; 'an Internet organ,' says Stelarc.

The Eye

The French philosopher René Descartes spent the most fruitful period of his life in the Dutch Republic, where he moved frequently between the academic centres – Franeker, Dordrecht, Leiden, Utrecht – developing his knowledge of mathematics, physics and physiology, before settling down in the remote seaside village of Egmond-Binnen to write out his new theory of everything. In 1632, he was to be found in Amsterdam; it is quite possible that he was among the audience at Dr Tulp's anatomy demonstration.

Descartes was no armchair philosopher. His radical ideas about the human body as a kind of mechanism, and the intellectual rigour that would soon turn him into an adjective – Cartesian – were based on first-hand observation and his own experiments. On one particular occasion during the 1630s, he took the unusual step of procuring the eye of an ox in order to understand more precisely the intricacies of vision.

He described his results in an essay entitled *La Dioptrique*, a work that is neglected in comparison with the one that prefaced it, the illustrious *Discourse on the Method of Rightly Conducting One's Reason and Seeking Truth in the Sciences*, which includes the famous aphorism *cogito ergo sum* – I think therefore I am. Descartes published *La Dioptrique* in 1637 at Leiden, accompanied by two additional volumes on meteors and geometry, making three major parts (and the preface) of what was to have constituted a grand *Treatise on the World*, other chapters of which he was forced to withhold when they were suddenly rendered obsolete by Galileo's discovery that the earth revolves around the sun.

He begins his discussion of the eye with an optical diagram. 'If it were possible to cut the eye in half without spilling the liquids with which it is filled, or any of its parts moving about, and the

plane of the cross-section passing right through the middle of the
pupil, it would appear as it is shown in this figure,' he tells us.
Already, it is interesting, in view of his philosophical subject matter, to note that Descartes is drawing our attention to something
that, for the very good reasons he gives, could not actually be seen.
The text that follows describes each of the parts that Descartes has
labelled in his diagram – the hard outer skin of the eye, a loose
second skin 'hung like a tapestry' inside the first, the optic nerve
and its branches, which mingle with fine veins and arteries across
the inner hemisphere of the eye in a fleshy third layer, and three
zones of different transparent 'slimes or humours' filling the interior of the orb.

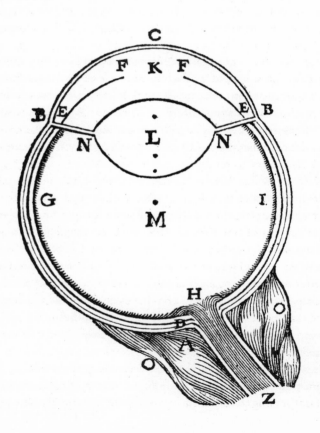

How do these liquids and nerves enable us to see? Descartes took up his ox eye and a scalpel and carefully peeled away the outer layers from the back until it was transparent. Then, he positioned the eye facing outward in a hole at a window in a darkened room. He placed a thin piece of white eggshell at the back, where he had cleared the surface. The bright scene outside was faithfully reproduced upside down and in miniature on the eggshell screen. As he wrote in his account of the experiment: 'you will see there, not perhaps without admiration and delight, a picture [*peinture*] which will represent in a strictly artless way in perspective all the objects which are outside.'

This internal image is formed by optical refraction, which explains why it appears upside down. The image of ourselves that we see when we look closely into the black centre of somebody's eye, on the other hand, is formed by reflection. This little effigy has inspired our word pupil, which derives from the Latin *pupilla*, meaning a little doll, as well as the charming seventeenth-century colloquialism 'to look babies at' somebody, which means to look adoringly into their eyes, a reference not as you might think to the urge to procreate with that person, but to the sight of this diminutive human form. This part of the eye was not called the pupil until the 1660s, however; Descartes describes it using the French word *prunelle*, meaning a sloe.

I decide to try to repeat the experiment. My prize-winning local butcher, Crawford White, is very tolerant of my occasional odd requests, and does not turn a hair when I ask for eyes of a bull or cow, although he explains he cannot get these for me, presumably because of the dangers of bovine spongiform encephalopathy. But, he tells me, he could have some pigs' eyes ready for me if I come back later in the day. At home, I gingerly open the little bag I have been given and find four pairs of eyes rolling around inside. Each eye is about the size of a grape, rather smaller than a bovine eye, which I worry may make the dissection tricky. Three-quarters of the spherical surface of the eye is covered by a white layer like a large icecap. The stump of the severed optic nerve protrudes from the midst of this expanse. The front of the eye is clear with glossy black and grey depths.

I pick out one of the eyes and begin by trimming away the flesh and fat that clings to it. Then, squeezing the eye slightly between my fingers in order to give it a firm surface, I begin to cut into the white membrane that protects the clear orb within. It is very tough, and I am nervous of applying too much pressure and piercing the inner membrane with my scalpel. Almost immediately, the worst happens and a gelatinous liquid oozes out of the eye. I take a second eye and start again. The same thing happens. I change my tactics, and try to shave the white tissue away rather than cutting through it. This works better, and by the fourth attempt, I have scraped away enough at the back of the eye that I can just perceive light through the remaining film.

At last, I take the eye over to a large cardboard box that I have prepared. I have cut an eye-sized hole into the front. At the back, I have cut out the shape of an upward-pointing triangle and positioned a bright light beyond it outside the box. I position the eye in the hole 'looking' through the box towards the light, and then I position my own eye directly behind, looking at it. I am thrilled to see a hazy image of the triangle pointing downward projected onto the white film.

'Now, having thus seen this picture in the eye of a dead animal, and having considered the causes, one cannot doubt that it forms a similar whole in that of a living man.' The eye, Descartes had discovered, works like a camera obscura, projecting an inverted image of the outside world on to its back surface. In his *Dioptrique*, he provides a ray diagram showing how this happens. It is both clearer and more beautiful than the diagrams I find in those few of today's anatomy textbooks that concern themselves at all with the physics of how the body actually works. In another version of the diagram, Descartes's illustrator has drawn the small bearded head of a man in miniature looking up at the back of the eye where the inverted image is formed. He seems like an astronomer gazing at the heavens.

This idea of a homunculus standing at the back of the eye sets up a paradox. For what does this little man gaze *with* other than *his* own eyes? Does the soul have eyes as well as the man? It is, says Descartes, 'as if there were yet other eyes in our brain'. Somehow, anyway, this image is converted into a form that can be transmitted through the brain to the soul, which Descartes locates in the pineal gland. This pea-sized organ is now known to be responsible for the release of the sleep-promoting hormone, melatonin. This gland is indeed sensitive to light, as the release of melatonin is triggered by darkness, but it is not in fact involved in visual perception.

Descartes's picture of the eye was incomplete as well as flawed – it offered no explanation of our ability to judge the size of things that we gain from having our two eyes spaced apart, for example. But it was revolutionary because it appeared to bring sight – the most mysterious, even mystical, of our senses, linked after all with 'visions' as well as straightforward seeing – within the scope of a mechanistic view of the body. Touch, taste and smell involve our physical interaction with the substance of the world. Even hearing, through the time it takes for a sound to reach us, is easy to imagine as some *thing* arriving at our ears from a distant source. Now sight could be comprehended in a similar way.

As I have pigs' eyes to spare, I decide to round off my experiment by trying to view an eye in cross-section, producing in reality the diagram that Descartes warns cannot be seen, by taking a bold slice through its equator. I approach the task with trepidation and some sense of horror. Luis Buñuel's image of a man slicing into a woman's eye with a cut-throat razor flashes through my head, even though I have never seen the surrealist film from which it comes. (Like Descartes, Buñuel actually used a calf's eye, as is all too apparent in the shot when I finally see it.) In the moment before I wield the scalpel, I understand why it is that organ donors are often more reluctant to let go of their eyes than they are even their hearts.

Yet when I actually cut into the eye, my perception changes. My knife is not sharp enough, and I cannot help squashing the eye out of shape as I depress the blade, spilling the contents. The worst over, my horror dissipates, to be replaced by fascination. Although they have not held in their correct positions, I can see that there are three distinct transparent liquids: a small quantity of a watery liquid, a larger quantity of liquid like jelly that has not quite set, and, slipping out from between them, a clear bead about the size of a pea. Though soft, it holds a definite shape that is oblate, and flatter on one side than the other. These are the aqueous humour, the vitreous humour and the lens, whose different refractive indices allow us to focus images of the outside world. Animal viscera are revealed as pure Cartesian mechanism. What began as an anatomical investigation has ended as a physics experiment.

Eyes are an important element of our identity. They are said to be the window on the soul. Even in fables of werewolves, the transformed man retains his human eyes. Yet what is it about them that conveys individuality? Colour is their most distinctive attribute. Eye colour was a feature of Alphonse Bertillon's system of identification for the Paris police, and has been routinely included in official identity records since the introduction of standard passports, where it supplements the likeness offered by a black-and-white photograph. The popular idea that eye colour is important seems likely to be reinforced once again with the introduction of iris-scanning technology to replace document checks.

This is doubly ironic because colour is not what is scanned. Iris scanners in fact use infrared light to detect unique *patterns* in the iris. And, although the iris is named after the Greek word for rainbow, it may come as a surprise to learn that there is no distinctive colour present in the eye in any case. The colours that we perceive do not arise from different pigments, but are what is known as 'structural colour' – an illusion of colour produced by an effect of interference between light rays that is also found in butterfly wings and iridescent bird feathers. All eyes contain a certain amount of one pigment, melanin. I found dark flecks of this pigment floating in the humours when I cut into my pig's eye. It is the variation in the levels of this pigment, together with the light-interference effect, that gives rise to the entire range of eye colours that we cherish. With progressively less melanin present, the eye can appear dark or light brown, hazel, green, grey or blue.

Francis Galton was curious to learn what eye colour had to say about heredity. He built himself a travelling case with sixteen numbered glass eyes of different colours. The eyes were set into a sheet of metal moulded in such a way as to give each of them eyelids and an eyebrow, an alarming surrealist touch when you first open up the case. Galton needed to be sure that the colour labels he chose from among the 'great variety of terms' used by compilers of family records were the ones important in nature. He didn't choose brown or blue, as we usually do, but categories of light and dark, splitting those with 'hazel' eyes into both camps. He then compared children with their parents and grandparents, whipping up his usual storm of statistics, but finding nothing more noteworthy to say at the end of it than that both blue eyes and brown eyes are observed to persist down the generations.

The ultimate answer to Galton's question about heredity came in 2008, when a team of (largely blue-eyed) researchers at the University of Copenhagen discovered a mutation of a particular gene that regulates a protein needed to produce melanin. Babies are often blue-eyed at first, even when born to brown-eyed parents, because this protein is yet to be released to its full extent. According to Hans Eiberg, who led the research, his genetic discovery suggests that all blue-eyed individuals alive today can trace their ancestry to one original Ol' Blue Eyes, who was the first to undergo this mutation, between 6,000 and 10,000 years ago.

A mere accident in nature, maybe eye colour doesn't have quite the significance we thought in culture either. Becky Sharp in *Vanity Fair* has green eyes, Anna Karenina has grey, James Bond, blue. The worse the novel, it seems, the more important it is to be exact in description. To wit, Judith Krantz's *Princess Daisy* has 'dark eyes, not quite black, but the colour of the innermost heart of a giant purple pansy'. But many of the most famous characters in fiction prove surprisingly elusive when it comes to eye colour. Mr D'Arcy merely thinks Elizabeth Bennet has 'fine eyes' in *Pride and Prejudice*. Julian Barnes devotes large portions of his novel *Flaubert's Parrot* to the matter of Emma Bovary's eyes, berating a (real-life) critic who had triumphantly spotted Flaubert's supposed sloppiness in describing her eyes variously as blue, black and brown. It doesn't matter, Barnes suggests; or rather it does, but not in the sense that we must know the colour of her eyes in order to identify, or identify with, the heroine. Emma's eyes are whatever colour Flaubert chooses to make them for reasons of his own at that point in the narrative. In *Tess of the D'Urbervilles*, Thomas

Hardy also dodges the question of the colour of his heroine's eyes, which are 'neither black nor blue nor grey nor violet; rather all those shades together, and a hundred others, which could be seen if one looked into their irises – shade behind shade – tint beyond tint – round depths that had no bottom; an almost typical woman'. If an author wants us to believe that his character is everywoman, then being vague about eye colour is a good way to start.

It seems that our sense of sight has grown in importance during human evolution, and this growth may be at the expense of other senses. For example, we have many genes involved in processing smells, but they are underused in comparison with the relatively few we have dedicated to vision. As sight has become more important to us, it is the brain's ability to process visual signals that has developed fastest. Our eyes themselves have not kept pace with our thirst for visual information, which may help to explain why, in a world where visual communication is increasingly important, so many of us nevertheless need to wear glasses.

In order to understand the extent to which vision is realized in the brain rather than the eye itself, and to which it overlaps with other sensory information, I pay a visit to the Cross-Modal Research Laboratory at the University of Oxford. The tiny laboratory resembles something between a toy store and a corner shop, stacked with odd gadgets and familiar food brands. Its director is Charles Spence, a professor of experimental psychology. He is wearing his trademark red trousers when I meet him, and speaks with an unnerving staccato delivery. The senses – the familiar five of sight, hearing, touch, taste and smell, though many more according to some beliefs – are usually considered in isolation, he explains. But we use them in concert. This leads to some very odd perceptions with disturbing implications. For example, Charles tells me, an interviewer is more likely to regard a job applicant as a serious candidate if the interviewer is holding a heavy file on his lap than if he is holding a lighter one. The weight of the file counts for more than what he sees and hears. 'Never mind the quality, feel the width,' it seems, is not just a desperate sales pitch, but an axiom in nature too.

Our unconscious mixing of sense signals can easily mislead us. It can also be exploited in order to alter our behaviour. Much of Charles's work is for product manufacturers who can make good use of multisensory discoveries such as the fact that the sound you hear as your teeth crunch on a crisp, and even just the rustling of the packet, is a significant factor in your perception of its flavour. 'We're interested in the interaction of the senses, both at the level of the single cell and how it comes together in the brain. Can you "taste the weight", for example? Or, how does the fragrance somebody is wearing affect your estimation of their age?'

Vision is surprisingly easy to fool, perhaps because our brains are so biased in favour of this sense. One famous experiment is known as the rubber-hand illusion. A subject's hand is positioned out of their sight, while an artificial hand (a rubber glove will do) is placed in the line of sight where they might normally expect their real hand to be. The experimenter then touches both the invisible real hand and the visible artificial hand with a synchronized stroking action. After a while, the subject begins to feel that the fake hand is really their own. A cruel extension of the experiment involves bringing a hammer down on the artificial hand: the subject cannot help flinching. In these situations, the brain is prioritizing visual information over weaker signals sent from receptors under the skin of proprioception, our sense of our position in space. The hand must be a reasonable likeness for the trick to work: a left glove for a right hand will not produce the effect. However, since a bright yellow rubber glove works perfectly well, it seems that skin colour for once doesn't matter.

A still more dramatic illustration comes from the psychologist Richard Gregory, who witnessed the recovery of a man who had been blind from birth until he was given a corneal graft. Gregory took the man to various stimulating venues in London, including the zoo and museums. At the Science Museum, he was shown a lathe, as he had always been interested in machinery. In its glass case, he was unable to recognize it. But once he had run his hands over it, he understood it fully. As Gregory tells it, 'he stood back a little and opened his eyes and said: "Now that I've felt it I can see."' The

moment explained why, on the journey to London, the subject had
been entirely nonplussed by the sights streaming by the car window.
The fact that he remained effectively blind to objects until he had
touched them indicates that neural pathways concerned with vision
had been taken over in his blindness by touch, and that his brain was
only now beginning to rewire itself.

Understanding how our senses overlap in the brain can lead to better treatments for sensory loss. For example, therapeutic procedures
using mirrors can help amputees who experience pain associated
with their lost 'phantom' limb and stroke victims who have lost
motor control on one side of the body by enabling them to compare
sensory feedback obtained by proprioception with what they see in
the mirror. One sense can even begin to replace another on a permanent basis. Blind people who use part of their brain normally
dedicated to vision to interpret the letters of Braille sometimes find
that the tactile sensitivity of the fingers is increased, giving them better spatial discrimination. In 1969, Paul Bach-y-Rita at the University
of Wisconsin in Madison scaled up this idea to create prosthetic 'eyes'
using arrays of vibrating pins acting like pixels to create crude images
of scenes recorded by a camera. The device, called BrainPort, was
initially designed as a vest to be strapped to the stomach, where the
large expanse of skin would serve as a touch-sensitive screen. Later
versions were miniaturized to fit on the surface of the tongue, which
is much more touch-sensitive. Bach-y-Rita's subsequent innovations
show that other senses may be recreated in the same way, such as balance in subjects who have suffered damage to the part of the ear
normally responsible for providing this sense. After a short period
using the BrainPort, modified to detect tilt, some patients even found
some restoration of 'balance memory' that lasted for several hours
after the device was removed. People learn to use such equipment by
a laborious process of conscious sensory translation, but as they
become more familiar with it, the brain's neural pathways adapt so
that the substitute sense is experienced more like the sense that has
been lost.

We are inherently multisensory beings. We see and hear together.
We use our senses of smell and taste together. Combined sense signals

often amount to more than the sum of their parts, and are more memorable. I am sure I would not recall a particular occasion when I was listening to the gods' entry into Valhalla from Wagner's opera *Das Rheingold* on the car radio if I had not been driving across the Severn bridge at that very moment, for example. It is only when he actually smells and tastes the famous madeleine that Marcel Proust's memories of lost time are unleashed; the sight of it alone is not enough to do this. The converse is true, too: take away one sense, perhaps one we don't even realize we are using at the time, and our perception is disproportionately impaired. A loss of sense of smell takes away much of the enjoyment of food, because so much of what we think of as taste is in fact linked to smell. Or, as Charles Spence's tests have shown, it may be important that a warning signal on a car dashboard is delivered by visible and audible means together, such as a flashing light with an intermittent tone. The brain may miss either of these signals on its own, but has a much better chance of registering the correlated event.

I ask Charles about synaesthesia, an effect that has always intrigued me in which a signal in one sense also stimulates a brain response in another. A synaesthete might find that musical tones correspond to certain colours and textures, or that shapes conjure up tastes, for example. Some of my favourite composers and artists have claimed synaesthetic experience: Kandinsky, Hockney, Messaien, Sibelius and F. T. Marinetti, whose *Futurist Cookbook* includes recipes that require the diner to eat with one hand while stroking silk or sandpaper with the other, or to eat in a flight simulator so that the vibrations stimulate the taste-buds. One of the more persuasive claimants is the author Vladimir Nabokov, who lists the colours he associates with the letters of the alphabet in his autobiography, *Speak, Memory*. For him, each letter retains its distinct colour when placed next to others in a word, unless it produces a diphthong which happens to exist as a single letter in another language (such as happens with *sh* and *ch* and other combinations in the Russian alphabet of Nabokov's early years); in this case, the single colour associated with the letter in the other language bizarrely stains the English letters that make up the equivalent phoneme.

Synaesthesia first attracted the attention of scientists in the late nineteenth century, when the multisensory experiences offered by the 'total artworks' of Wagner, Post-Impressionism, absinthe and opium doubtless encouraged them on their way. Little progress was made in understanding it, however, owing to the highly subjective nature of the phenomenon. It is only now being dusted off by neuroscientists, who are interested in what it might tell us about the brain's ability to cross-wire the senses.

What is synaesthesia? Is it a condition, a delusion, an advantage or a curse? It is not listed in the most widely used psychiatric manual. It seems it is a neurological state of being, but not a neurological disorder, less like a condition and more like a cartoon hero's 'super power'. It's not quite all wine and roses: 'Synaesthetes may not be able to read a book because of the rush of extra information,' Charles tells me. But they do enjoy the extra sensations and are better able to remember things as a result. 'Synaesthetes wouldn't take a pill to get rid of it.' Synaesthetes come across rather like the members of an exclusive arty club. Certainly, there are many who clamour for admission. Conspicuous aesthetes such as Rimbaud and Baudelaire wrote works implying they had synaesthesia, for example, but scholars now believe that their experience of it was entirely vicarious, and they probably only picked up on the concept from medical reports. In modern tests, it is found in fact that women are more likely to experience synaesthesia than these male pseuds. Yet any of us might refer without affectation to a flower bed being a symphony of colour. We know there is music called the blues. Perhaps we are all latent synaesthetes.

Even blind people can have synaesthetic visual experiences, for example 'seeing' flashes of colour in response to aural signals such as numbers or letters being read out. It is thought that infants may possess neural pathways crossing over between the aural and the visual that are cut off in later life. One neuroscientist, Vilayanur Ramachandran, who has also worked on phantom limbs, describes the extraordinary case of one blind patient who began to notice that whenever he touched objects or read Braille his mind would conjure up flashes of light or vivid images (although not pictures of the item being touched). Such experiences suggest that neural pathways to the brain behind the damaged eye are somehow commandeered by these aural and tactile signals. Other blind people find that their hearing improves. It may not be their acuity to any sound that increases, but the ability specifically to process sounds that assist with spatial perception, which becomes such a challenge with the loss of sight. Individual cases such as these provide growing evidence that what scientists term the 'neuroplasticity' of the brain is directed towards restoring or enhancing its useful function when needed rather than merely shuffling the senses around at random.

The Stomach

It is not clear at what point the Very Reverend Doctor William Buckland's project to eat everything tipped from earnest scientific enquiry to sheer silliness.

Buckland was a noted geologist, and the first professor in the subject at the University of Oxford. His masterpiece was *Vindiciae Geologiae*, a work which set out a new theory for the subject in which fossils predated Noah's flood, but the Bible nonetheless remained literally true – a cunning piece of interpretation that relied on defining the 'beginning' in Genesis as some vague time after the formation of the earth but before the advent of man and other present-day species. He was the first to identify coprolites – fossilized dung – which provide our only direct evidence of what the dinosaurs ate. Against a background of suddenly rising grain prices, he wrote in favour of scientific agriculture, proper land drainage and irrigation, and allotments for the 'labouring poor'. All this work secured Buckland's reputation, and he rose in academia and the church to become the canon of Christ Church, Oxford, and later the dean of Westminster.

Buckland was not a man to be confined by establishment conventions, however. At the same time as pleasing everybody with his efforts to square the story of the Bible with the evidence of geology, he pursued a plan to taste the flesh of every animal. Though a notable scientist, he seems to have left no systematic record of his extended gustatory experiment; what information we have comes from anecdotes that continue to be passed down by Buckland's descendants. It might be supposed that the project was an extension of his wish to identify new sources of food for the growing population, but in reality it seems to have been simply his own eccentric whim. He tried hedgehog, crocodile, panther, puppy and garden snail. The naturalist Richard Owen was entertained to roast ostrich, which he found tasted like 'coarse turkey'. The critic John Ruskin much regretted missing another dinner at which was served 'a delicate toast of mice'.

Certainly, Buckland was not afraid to shock his contemporaries. Visiting a cathedral overseas, his attention was drawn to 'a martyr's blood – dark spots on the pavement ever fresh and ineradicable'. His scepticism aroused, Buckland knelt and put his tongue to the spots, and was immediately able to report otherwise: 'I can tell you what it is; it is bat's urine.' The account of one exceptional meal only emerged nearly fifty years after his death, when the writer Augustus Hare recollected being at a dinner with Lady Lyndhurst at her house in Nuneham near Oxford, where the heart of a French king (perhaps that of Louis XIV or Louis XVI; accounts differ) was said to lie preserved in a silver casket. As Hare tells it:

> Dr. Buckland, whilst looking at it, exclaimed, 'I have eaten many strange things, but have never eaten the heart of a king before,' and, before anyone could hinder him, he had gobbled it up, and the precious relic was lost for ever. Dr. Buckland used to say that he had eaten his way through the whole animal creation, and that the worst thing was a mole – that was utterly horrible.

In a footnote, Hare adds: 'Dr. Buckland afterwards told Lady Lyndhurst that there was one thing even worse than a mole, and that was a blue-bottle fly.'

Buckland's unusual hobby did nothing to hinder his advance. Perhaps it even helped: as dean of Westminster in 1845, he used his position to improve the meals for boys at Westminster School – who knows what treats they were served. He died in 1856 aged seventy-three from an infection of the spine which spread to the brain, so we can presume at least that nothing he had eaten did him any lasting harm.

Of course, items such as ostrich and crocodile are now available in our fancier food emporia. My well-thumbed copy of *The Joy of Cooking*, an American classic, includes instructions for cooking porcupine, raccoon, bear ('Bear cub will need about 2½ hours' cooking; for an older animal, allow 3½ to 4 hours.') and other road-kill recipes.

What does this unlucky bestiary say about the human stomach?

Buckland seems mainly to have eaten whatever came his way partly out of a desire to amuse or shock. The reverend doctor certainly had no scriptural qualms concerning the prohibitions of Leviticus, since he doubtless ate numerous 'abominable things', including the blue-bottle, which falls into the prohibited category of winged insects. If he did not try other unclean species such as hoopoe and hyrax, it is probably only because the opportunity never presented itself. Look-ing at his project more broadly, it seems reasonable to suppose that if he had put a fraction of the effort he put in to eating animals towards the evaluation of new vegetable sources of nutrition, he might have made a more practical contribution to the problems of feeding the world's population.

The catalogue of animals that Buckland consumed is perhaps chiefly remarkable for reminding the rest of us how very few species generally make up our diet. The stomach is one of the simplest parts of the body to have earned the name of organ. As I am forced to acknowledge on seeing one for the first time in the dissection room, it is really just a bag, and as with any bag, you can put anything into it as long as it fits. Chevalier Jackson, a Philadelphia laryngologist, made a collection of the objects he retrieved from patients' throats and stomachs 100 years ago. He amassed thousands of items, includ-ing keys, a padlock, nails and unclosed safety pins. The Gordon Museum, which houses the pathological anatomy collections of a number of the London teaching hospitals, exhibits a similarly extreme miscellany of objects swallowed down the years on purpose or by accident, including the bedsprings found in the stomach of a Brixton prison inmate who had been that desperate 'to get out for a while'.

Humans are natural omnivores. Although we are not equipped with sharp teeth and claws and the speed to catch prey, we do have the big brains conferred on us by evolution which have enabled us to use tools and cunning to diversify our diet. Our baglike stomach can hold anything, and the five metres of intestines that lie downstream of it can make a good job of digesting most things. We can digest raw meat, but the discovery of fire gave us the ability to process meat far more effectively, and so to eat far more of it (more than is good for us). On the other hand, we can digest surprisingly little of what the vegetable kingdom has to offer, preferring as we do ripe fruit to grass and bark, since our stomachs lack the compartments that act like fermentation vats in true herbivores, breaking down the more fibrous organic matter. The disgust that we feel, therefore, when we are forced to consider a meal of bluebottle, or for that matter the human consumption of a human heart, is based entirely on culture rather than nature.

In perhaps the most famous of his essays, 'On the Cannibals', Michel de Montaigne wrote: 'I think there is more barbarity . . . in lacerating by rack and torture a body still fully able to feel things . . . than in roasting him and eating him after his death.' And indeed, human flesh, like most flesh, is nourishing to the omnivorous human body. What does it taste like? 'Human flesh tastes like pork,' according to Helen Tiffin, the author of an essay exploring humanity's debt to the pigs that feed us and that now grow our replacement organs, too; 'hence', she continues, 'the term "long pig" for human meat cuts, the "long" denoting the difference between the limb lengths of pig and human. Although there are few "first hand" accounts of the flavor of human flesh, its similarity in texture and taste to that of pork seems generally agreed upon.' If the use of cadavers in medicine is acceptable utilitarian practice, Montaigne reasoned, then why isn't their use for nourishment? In an age when physicians would taste a patient's blood as an aid to diagnosis, and ground human skull (with or without ginger) was prescribed as a remedy for fits, cannibalistic practices might even be regarded as a legitimate medicinal use for the body.

Although tales of cannibalism continue to excite us as they once did Montaigne and Defoe and Melville, anthropologists had more or less set it aside as a topic for serious study. Alleged instances were historical or otherwise hard to confirm, and the sensationalism aroused was giving anthropology as a whole a bad name. However, interest was revived by an outbreak of the prion disease kuru in the mid twentieth century among the Fore people of the Papua New Guinea highlands. Prions are infectious agents based on proteins rather than on nucleic acids, as viruses and bacteria are. Prion diseases cause progressive loss of muscular coordination which is typically characterized by trembling, dementia and paralysis. The Papua New Guinea epidemic ultimately killed more than 2,500 people. The peculiar pattern of incidence of kuru was supposedly explained by a residual custom of cannibalism as the mode of transmission. Women would eat the brain and spinal cord of deceased relatives and become infected, as would children through contact with their mothers during ritual feasts. Men, who ate primarily the less infective muscle tissue, showed a lower infection rate.

But the American anthropologist William Arens is sceptical in regard to all claimed instances of ritual cannibalism, including even these ones where anthropological study is supplemented by medical scientific investigation. Arens notes that both medical and social scientists have tended to take unconfirmed stories of cannibalism as read, and that even professional anthropologist 'witnesses' may have been duped by seeing natives consuming what is in fact pig meat. (Melville teases the reader with just such a confusion in his South Seas novel *Typee*, where the mere lighting of a fire is enough to excite the fears of the two castaways at the centre of the story that they are for the pot. Then 'some kind of steaming meat' is brought out: 'A baked baby, I dare say!' They find it tastes 'excellently good . . . very much like veal'. But their terror is revived when they recall that there are no cows on the island. 'What a sensation in the abdominal region! Sure enough, where could the fiends incarnate have obtained meat?' Finally, one of the men holds a lighted taper over the pot and to his great relief identifies 'the mutilated remains of a juvenile porker'.) When the infectious protein or prion responsible for kuru was duly traced, the cannibalism route for transmission remained largely accepted, although in Arens's view the evidence was still only 'circumstantial'. How is it, he wonders, that cannibalism explains kuru seen among a remote New Guinean tribe, but is not suggested as a route for the transmission of the related Creutzfeldt-Jakob disease when it is found among developed societies? In general, controversy persists about the very existence of ritual cannibalism, with no solid evidence of contemporary practice and only unsubstantiated evidence of past practice. But if the practice is absent, the fear of it is apparently universal. More detached anthropologists have observed that very often the 'primitive' groups under investigation for supposed cannibalistic behaviour turn out to have their very own cannibalism myths about those who have come to study them!

From variety to quality – and quantity.

The potential for the human stomach to accommodate, if not everything, then at least an exciting variety of what nature has to offer leads us to the idea of the delicious. Why, when so much can be called food, are we so picky about what we eat? It is usually wise to turn to the French for advice in these matters, and none stands higher in this field than the author of the quasi-scientific masterpiece *The Physiology of Taste*, Jean Anthelme Brillat-Savarin, the only man, so far as I can discover, to enjoy the uniquely Gallic honour of having a cheese named after him. 'To eat is a necessity,' he observed. 'To eat well is an art.'

Published in 1825, when Buckland was only on the hors d'œuvres of his grotesque eating experiment, *The Physiology of Taste* set the parameters for an emerging national cuisine in post-revolutionary France. It is a glorious mélange of recipes, history, humorous anecdotes, invented words, autobiography and food science, all prefaced by a number of 'aphorisms of the professor', which include the still-famous comments 'Tell me what you eat, and I shall tell you what you are,' and 'The discovery of a new dish does more for human happiness than the discovery of a star.' Brillat-Savarin, who was a lawyer and wrote large portions of the book in dull interludes while sitting in court as a judge, is accurate enough for his time on the workings of the human sense of taste, and why some of us are non-tasters while others are super-tasters, able perhaps to discern 'the latitude under which a wine has ripened' or 'the special flavor of the leg upon which a sleeping pheasant rests his weight'. He explores the close relation of taste and smell, drawing much from an interview with a man whose tongue has been cut out as a punishment. He's good, too, on the patent inadequacy of our convention of breaking down flavours into sweet, sour, bitter and salt. Modern classifications also admit the sensation of hotness imparted by chillies and an aromatic savoury sense, *umami*, from the Japanese, which Brillat seems to anticipate with his coinage of the term 'osmazome' to describe the depth of a good stock. But these few terms do not begin to compass the infinity of tastes, which would need 'mountains of folio foolscap to define them, and unknown numerical characters for their classification'.

He has less to say, however, on how the gut digests food and the ways in which we extract from it energy, proteins, vitamins and minerals. For Brillat's chief concern is our pleasure in food, as his book's subtitle reveals: *Meditations on Transcendental Gastronomy*. In short, he wants us all to be gourmands. 'Gourmandism is an impassioned, considered, and habitual preference for whatever pleases the taste,' he writes – and not to be confused with gluttony, he adds hastily, for gourmandism 'is the enemy of overindulgence'. He identifies some natural classes of gourmand: the clergy, writers, bankers, and also doctors, whom he nevertheless chides for their unpleasant-tasting medicines and austere dietary regimens. Gourmandism is for girls, too: 'it is basically favorable to their beauty'. Gourmands make better marriages and live longer.

Yet the pleasures of the table – humankind's legitimate recompense for being the only species to experience suffering, says Brillat – are too easily taken to excess. Even the French aren't so fastidious about food that they can entirely avoid gratuitous overeating. Clerics are especially prone. Rabelais, in the fourth book of *Pantagruel*, attacks monks as men who profanely make a god of the belly. These 'lazy, great-gulletted Gastrolaters', as he calls them, worship the god Gaster and bring him offerings – Rabelais's list runs to several lip-smacking pages of meat dishes, with several pages more given over to fish dishes that might be presented on fast days, when it was necessary to abstain from flesh. But Gaster is unimpressed with it all, and rudely points the idolaters to his lavatory, 'there to ponder, meditate and reflect upon what godhead they found in his faeces'.

Brillat-Savarin anticipated the problem of excess, and surprisingly includes in his book a very contemporary chapter on obesity. He's not worried about himself: 'I have always looked on my paunch as a redoubtable enemy; I have conquered it and limited its outlines to the purely majestic,' he announces. (The *Physiology* was the work of a lifetime, and was published the year before Brillat died at the age of seventy, so it seems reasonable that he should have attained a certain girth.) But he is concerned about a group of people for whom he invents another new word, the 'gastrophores', those who, through excessive consumption of starch and sugar, lack of exercise and too much sleep, 'lose their form and their original harmonious proportions'. Brillat-Savarin's remedies spring obviously enough from his identification of these causes and still deserve consideration today. But if exercise and dietary self-discipline fail, he also has an 'antifat belt' to recommend, a restraint on the belly to be worn day and night. He warns against more extreme measures, such as the habit, then prevalent, of women drinking vinegar, telling the touching story of a girl he had known in his youth who wasted away from what we would now recognize as anorexia (the condition was named in the 1860s) and ultimately died in his arms. Thus, perhaps, did the happy marriage of the gourmand elude Jean Anthelme.

A notorious modern landmark in French gustatory excess is the film *La Grande Bouffe*, in which, over the course of a weekend, and attended by a number of prostitutes, four middle-aged men gather in a secluded villa and resolve to eat themselves to death. One by one, they succeed. *Fin*. It's the kind of film that gives European cinema a bad name in the Anglo-Saxon world, where any sort of carnal appetite is best ignored. *La Grande Bouffe* caused outrage when it was first shown, at the Cannes Film Festival in 1973, not so much because of its morbid stew of food, sex and death, but more for its Italian director, Marco Ferreri, who had dared to bring his satirical gaze to rest upon the central rite of French life.

At one stage, the four men challenge one another to see who can eat the fastest, anticipating one of today's odder public spectacles – extreme eating. In these modern contests, however, there is no art about the food, and no allegorical edifice of civilization to be demolished like a croquembouche. You just eat as much of one thing – peas, oysters, Mars bars, peanut-butter-and-jelly sandwiches – as you can. According to the International Federation of Competitive Eating, the champions of this dubious activity are termed – and this at least would please Rabelais and Brillat – gurgitators. One Patrick Berto-letti holds records for eating the most key lime pie, pickles and pizza slices, as well as 275 jalapeño peppers in eight minutes. It is a surprise to discover that champion gurgitators are not all fat. Sonya Thomas weighs just 105 pounds but still ate forty-four Maine lobsters in twelve minutes. It takes physical preparation and training to succeed at the highest level. Medical assistance is on hand at the contests, but then, I realize, so it is at major sporting events. Anatomical examin-ation of successful competitive eaters shows that they develop stomachs capable of stretching well beyond the usual limits. Other matters remain a mystery. In striking contrast to Ferreri's film, the International Federation of Competitive Eating does not dwell on how its gurgitators void the excess they have consumed, for example. Consumption is encouraged, but its concomitant waste is politely denied. The phenomenon is growing in popularity, drawing tele-vision coverage and corporate sponsors, which include the expected food companies, but also Procter & Gamble, the maker of Pepto-Bismol. 'Competitive eating', though only just beginning to be studied by human psychologists, is an established term used by ani-mal behaviourists. The terminology is a reminder that, quite aside from the public spectacle, these contests provide a grotesque epitome of nature's contest for the survival of the fittest.

In *La Grande Bouffe*, the last of the four men to go, Philippe (played by Philippe Noiret), feeds titbits to one of the dogs that have gathered in the garden of the house, looking forward to their own feast. 'Be greedy,' he advises the animals. 'Eat too much. Always eat too much.' His last act is to tuck in to a huge jelly in the shape of a pair of breasts, a symbolic return to his first ever meal. As Ferreri's film and today's eating contests each illustrate in their own uncompromising ways, the veil that the art of gastronomy draws over essential nourishment is as fragile as spun sugar.

The Hand

I want you to try this: hold up your left hand and pinch the thumb and forefinger to form an O. Then crook the remaining fingers at the middle knuckle, or what medical types in their helpful way call the proximal interphalangeal joint (the bones of the finger are called phalanxes). It is not especially easy. You will find you have an urge to bend the other joints too. But resist this (as much as you can), and instead hold your free fingers with their middle joints at as close to a right angle as you can manage and the other joints as straight as possible. This will feel a little odd. It is not a spontaneously adopted position, and it requires the use of certain muscles and not others in a way that is not usual.

This is the exact pose that Doctor Tulp is holding in Rembrandt's painting with which I began this book. Art historians and medical historians alike have examined the painting minutely, but they mostly pass over this detail. This is strange because it is clearly an important detail to the artist, who has placed tiny dabs of white paint on Tulp's manicured fingernails so that they catch the light. The only other thing that catches the light in this way is the shiny pair of forceps which Tulp grips in his right hand. Most scholars have assumed that Tulp is merely waving his free hand in a rhetorical gesture. One, however, William Schupbach, has observed that Tulp holds his hand and fingers in this peculiar way in order to demonstrate the action of the very muscles that he is simultaneously lifting from the dissected arm in front of him. Rembrandt thus presents him dutifully executing both of his functions as praelector of the Amsterdam surgeons' guild: he is both performing a dissection and giving a lesson about the workings of the human body. In a broadly humanist way, he is demonstrating the objective similarity of the dead and the living. We can presume that he is talking his attentive audience through what he is doing at the same time. In that case, notes Simon Schama, 'Tulp is seen at the precise moment of demonstrating two of the unique attributes of man: utterance and prehensile flexibility.' These, of course, make no humanistic point, but demonstrate rather man's God-given uniqueness.

When you bent your fingers a moment ago, you will have felt and perhaps seen a certain muscle contract in your forearm. This is the flexor digitorum superficialis, in other words the surface muscle that flexes the fingers. In the lower arm, it narrows and divides into four tendons that pass through the wrist. Each of these four tendons then forks in two when it reaches the end of its journey, and these pairs of tendon branches fasten to opposite sides of the middle joint of each finger. This bifurcation is an especially elegant piece of design as it allows for a second set of tendons, which run from a different flexor muscle, the flexor digitorum profundis, to pass through the gaps in order to operate the *terminal* joint on each finger. These eight tendons control the curling of the fingers like puppet strings. On the other side of the arm lie muscles called extensors from which run further tendons that serve to straighten each finger. In addition to a general extensor, there are individual extensors for the forefinger and the little finger, which explains why your forefinger is more effective as a pointer than the longer second finger, and why we are apt to breach English tea-drinking etiquette by leaving our little finger sticking out rudely in the air while the other fingers grip the cup handle (a gesture that may stem from chivalric etiquette when it was better to display refinement by not grasping everything edible with a greedy fist). All in all, says J. E. Gordon in his brilliant book *Structures*, these tendons 'run through the body in almost as complicated a way as the wires of an old-fashioned Victorian bell system'. The fingers themselves contain no muscles, and human dexterity is therefore achieved entirely through this marionette-like remote control. Rembrandt's and Tulp's opting to illustrate this aspect of the human anatomy allows them to make a revolutionary new point, the one shortly to be articulated in detail by René Descartes, that the body may be regarded as a kind of machine.

We have seen that a number of falsifications surround the dissected hand in *The Anatomy Lesson of Doctor Tulp*. It may have been painted from a separate anatomical specimen, and not even belong to Adriaen Adriaenszoon, the man on the slab. And since no real dissection ever started with the hand, it is likely that the artist and his patron agreed to focus on the hand for reasons to do with the beauty of its intricate anatomy and its indication of the divine in the human. Oddest of all, though, is that the muscles and tendons that Tulp is holding cannot be from a left arm at all. They lead from the wrong side of the elbow. It seems that Rembrandt must have worked from a right arm, and then replicated what he drew like a decal over the left arm of Adriaenszoon. The puzzle remains as to why the praelector agreed to let this travesty of his own art of surgery be set down for ever on canvas. Perhaps he was more concerned with his own portrayal.

Rembrandt's brushstrokes nevertheless show a very beautiful dissection, done to a higher standard than any of the hands I saw in modern dissection rooms, and easily the equal of those I found preserved in anatomical collections. It fully lives up to the contemporary view of the hand as man's noblest appendage. For Helkiah Crooke, writing 'On the excellency of the hands' in 1618, they are the 'two wondrous weapons' of man and no other animal. The hand is the 'first instrument so it is the framer, yea and imployer of all other instruments. For not being formed for any one particular use it was capeable of all . . . By the helpe of the hand Lawes are written, Temples built for the service of the maker, Ships, houses, instruments, and all kind of weapons are formed.'

This general versatility is the mark of the hand's – and our own – superiority. It is not the specialized claw of less able creatures. Especially with the extension of tools, the hand is capable of all things. It becomes the physical analogue of our free-roving mind. The pre-Socratic philosopher Anaxagoras believed that man was more intelligent than the beasts because of his hands. Aristotle, about a century later, believed roughly the opposite, that our hands only became necessary with intelligence. Either way, they were in agreement that manual dexterity and intelligence are closely connected. This remains the consensus today, although which came first is still hotly disputed.

The apparently simple act of pointing with the index finger shows how closely bound up the use of the hand is with the development of other human abilities. Helkiah Crooke and others thought we were the only animal to use tools. This belief has been disproved by observations of chimpanzees and certain other species, which takes some of the edge off the historic case for the uniqueness of the human hand. However, we remain, so far as we know, the only creature that points. Pointing is a highly 'unnatural' action. To point at something presupposes that we have a mental label or name for what is being pointed at, or else the action would mean nothing. This in turn requires the existence not just of language but of a shared language and, moreover, our understanding that the person for whom we are pointing has a mind similar to our own, so that they can infer exactly what it is, of the many things that may lie in front of our finger, we are pointing at. According to the physician and philosopher Raymond Tallis, this makes pointing 'a fundamental action of world-sharing, of making a world-in-common'.

The pointing hand soon acquired a life of its own, known as an index, fist or 'manicule'. Henry VIII drew his own finely inked pointing hand symbols in the margins of his books when he wished to be able to find certain passages again. Manicules were often beautifully drawn in highly individual styles, reinforcing the sense that they were not merely markers but heartfelt personal gestures. The pointing hand became one of the first clichés – clichés originally being the special symbols printers needed so often that it was worth casting a special piece of type for them. Regarded as a standard punctuation mark until the eighteenth century, the manicule was revived in the 1980s as the user-friendly cursor symbol on computer screens. Lone hands may point and perform other helpful duties, like Thing, the severed hand-servant in the Addams Family cartoons, who lights Gomez's cigars. But they also point fatefully *at* us, like the evil glove of the Blue Meanies in the Beatles film *Yellow Submarine*, the airborne hand of the UK National Lottery 'It could be you' advertisements and the commanding fingers of General Kitchener and Uncle Sam in posters from the First World War.

Pointing is just one of a huge vocabulary of gestures of which the hand is capable. Indeed, it has been estimated that there are more possible hand gestures than there are words in the English language. The Hand of God does not only point, but also extends two fingers together (in blessing) and offers an open palm (spreading beneficence upon the earth). In 1644, John Bulwer, a man so obsessed with hands that he named his adopted daughter Chirothea (Hand of God), published *Chironomia* and *Chirologia*, an exhaustive catalogue of human gestures. Bulwer believed that gesture was based on 'universal reason', which was independent of language, and might be adopted as a kind of silent esperanto. He offers some insightful explanations for familiar gestures, such as:

> TO WRING THE HANDS is a naturall expression of excessive griefe, used by those who condole, bewaile, and lament. Of which Gesture that elegant Expositour of Nature hath assign'd this reason. Sorrow which diminisheth the body it affects, provokes by wringing of the minde, tears, the sad expressions of the eyes; which are produced and caused by the contraction of the spirits of the Braine, which contraction doth straine together the moisture of the Braine, constraining thereby tears into the eyes; from which compression of the Braine proceeds the HARD WRINGING OF THE HANDS, which is a Gesture of expression of moisture.

The length of his descriptions in these volumes lends some force to his argument in favour of a language based on gesture rather than words. It is a relief as well as a pleasure to turn to the tiny engravings, arranged in grids of twenty-four on a page, of hands upraised, drooping, clenched, outspread, drumming, stroking, clasping and waving, each with its own immediate expressive power.

Bulwer is stronger on gestures of devotion than he is on gestures of vulgar abuse, but many of the latter go back even further in history. In his play *The Clouds*, Aristophanes gives a stage direction for Strepsiades to give Socrates 'the finger' when Socrates asks him a question about beating out a rhythm. 'Why it's tapping time with *this* finger,' Strepsiades answers. 'Of course when I was a boy,' he adds, now raising his phallus, 'I used to make rhythm with this one.' The phallic connotation of the finger is unmistakable, and it would be no surprise to find that it extends back well before the ancient Greeks. Other gestures have largely escaped their obvious vulgar connotations. Both the upraised thumb and the 'okay' gesture of forefinger and thumb closed in a circle are positive signs for most of us, although in Greece and Brazil respectively they remain unspeakably offensive.

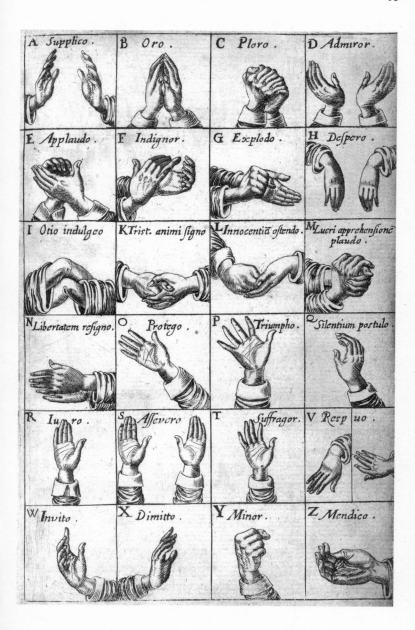

The origins of the British equivalent of the 'finger', the V sign, are more mysterious. One story is that English longbowmen, captured during the Hundred Years' War with France, would have their first two fingers – the ones that guide the flight of the arrow – cut off so that they would be useless when they returned to the battlefield. Bowmen who had never been captured would therefore wave their intact fingers at the enemy as a gesture of defiance. A V sign also makes an appearance in the elaborate armoury of hand signals described by Rabelais in an absurd duel in *Pantagruel*. The English Thaumast has come to Paris to learn from the wise giant Pantagruel, whereupon Pantagruel's mischievous companion Panurge waylays him with a gestural battle of wits. It is Panurge and not the Englishman who makes the V sign, though. So inventive and silly are the gestures that Rabelais describes that it is impossible to be sure whether this V sign has any particular meaning. Like Bulwer, Rabelais is keen to suggest the communicative power of hand signals, but he ends up demonstrating mainly that it is the coarsest gestures that are the most widely understood.

The most important way in which our hands have informed our intelligence is by giving us a readymade numbering system. The Roman numerals I, II, III and IIII may be based on the upheld fingers, with the symbol for five, V, based on the shape made by the thumb and the forefinger when the whole hand is held up. So-called 'denary' counting is based on the ten fingers and thumbs, and most other popular number bases, such as binary and bases four, twelve and twenty, are based on various combinations of limbs and digits. Even an octal system used by some Native American cultures begins with the hands: it counts not the peaks that our fingers make but the valleys in between them.

No animal has evolved with more than five digits since the terrestrial vertebrates (reptiles, birds and mammals) branched off on their own evolutionary path at the beginning of the carboniferous period 360 million years ago. But why do we have as many as five? As we have seen before, nature tends to furnish us with as many of the parts as we need, and where these are replicated, as with eyes and ears, it is for a good reason. So what do the five fingers do – either individually or together in various combinations – that gives them all a role?

In counting, each finger is exactly equivalent. But for most other tasks, they are as varied as the tools on a Swiss Army knife. The index finger is the best pointer because of its length and the dedicated muscle that straightens it. It is also more manoeuvrable than the other fingers. The index finger belonging to Inspector Bucket in Dickens's *Bleak House* is so versatile that it is almost a character in its own right, a confidant for Bucket as he puts it to his lips, to his ears, and rubs it over his nose before wagging it before a guilty man. 'The Augurs of the Detective Temple invariably predict that when Mr Bucket and that finger are in much conference, a terrible avenger will be heard of before long.'

The middle finger, though a little longer than the index, is a poor pointer: try it and you'll find that it is hard to straighten this finger while holding the others out of the way. But it has other uses. The Romans called it the digitus impudicus, the finger of impudence – perhaps they took over the custom of giving people 'the finger' from the Greeks. It is also called the digitus medicus because Roman physicians were apparently in the habit of using it to stir medicines. Next comes the digitus annularis, which we still call the ring finger, 'annulus' being the Latin for a little ring. It has this function for symbolic reasons rather than because it is especially adapted to the task. The ancients believed (incorrectly) that this finger was directly linked to the heart by a special vein. The little finger is called auricularis, because even it is not useless: it is just the right size for cleaning out the auricle of the ear.

Finally, there is the thumb, 'the father of technology', according to Raymond Tallis. It is our having an opposable thumb – meaning to say we can employ it in opposition to the other fingers – that greatly increases the capability of the hand, so that it is able to exert a wide variety of grips. Montaigne, in his note 'On thumbs' in the *Essays*, gives a correct derivation for the French for thumb, *pouce*, from the Latin *pollere*, meaning to excel in strength. He also offers a spurious but nonetheless apt alternative name, *anticheir* (derived from the Greek, it would mean 'opposed to the hand'); both terms are revealing about the unique importance of this digit.

But it is the opposed thumb together with the mutually independent operation of the fingers that really gives us our dexterity. The names of individual fingers hint at dedicated uses for each of them, but there is so much more they can do in their many permutations, from the way the forefinger and thumb delicately pinch in order to pick a flower or remove a contact lens, to the carefully balanced use of all five to wield a pair of chopsticks. Add to this the dizzying speed of other manipulations – of cards by a trickster, of strings by a guitarist – which earns its own special name of prestidigitation.

Palms have been 'read' for millennia, but the practice has only recently been put to the scientific test. The tradition was perhaps made respectable by Aristotle when he observed carelessly in his *Historia Animalium* that the life line across the hand seemed to be longer in long-lived persons. Why should it be the hand that reveals our fate? It seems simply that the palm of the hand contains a suitable number of legible features and is easily offered for inspection. In 1990, scientists at the Bristol Royal Infirmary looked at the life-line length of 100 consecutive autopsies. Surprisingly, perhaps, they did find a correlation between the length of the life line and the age of death. But it was not quite the evidence for the veracity of palmreading that it might seem. As the scientists point out: 'With increasing age we all become more wrinkly.' In their paper, they admit that a better idea would be to monitor the life-line length of a number of subjects over a lifetime, preferably, they add tongue-incheek, 'with investigators meeting every 10 years in exotic locations to report preliminary results'. Such a study has yet to be undertaken.

Another obvious feature of the hand is the length of the fingers in relation to one another. They were once thought to denote five (not Shakespeare's seven) ages of man, from the little finger of youth to the ring finger at the time of marriage to the longer fingers of maturity and finally to the decline of the thumb. In 1875, a German anatomist and anthropologist named Alexander Ecker observed that women's index fingers tend to be longer than their ring fingers, while men's are the other way round. This was such a curious finding that others rushed to confirm it – which they did, but when nobody could work out what it actually meant, the information was quietly ignored. This was the case until 1983, when Glenn Wilson at the Institute of Psychiatry in London responded to an invitation from the *Daily Express* to contribute to a survey of 'changing attitudes of women in the 1980s'. His questionnaire to female readers asked about their assertiveness and competitive instincts, and requested, by the by, that they measure the lengths of their fingers. The results showed some tendency for women with a low index to ring finger ratio to be more assertive, in other words the women with more male-like fingers behaved more like men (it having been taken as read, it seems, that assertiveness is essentially a male attribute). This discovery confirmed finger ratio as a convenient indicator of the level of testosterone to which a person has been exposed in the womb. Research based on finger ratios has mushroomed, and they have now been used in studies of sexual selection, sexual orientation, fertility, spatial reasoning, sporting ability, musical talent, autism (which occurs primarily in males), and success in financial trading. In 2010, investigators at the University of Warwick obtained results to suggest that men with long index fingers are less likely to get prostate cancer. It seems the hands have much to tell us yet.

The most pervasive and most divisive conclusion we have drawn from our hands is that one is better than the other. Preferment of the right hand develops very early in life. There is evidence that at fifteen weeks' gestation most of us exhibit a preference for sucking our right thumb. Tracking a number of subjects before and following birth, Peter Hepper at Queen's University Belfast found that the prenatal right-handers all kept their right-handedness as children. Most of the left-handers retained their preference for the left, but some switched to the right.

It clearly disturbs us on some deep level that we possess these two sets of limbs that appear to be entirely symmetric and yet are generally asymmetric in use. This imbalance is one of the oldest bases for discrimination: between left and right. The Bible repeatedly declares a bias in favour of God's (and everybody else's) right hand and against the left. Arbitrarily, it seems, in the gospel according to Matthew,

God demands that those 'on the left hand, Depart from me, ye cursed, into everlasting fire', while those on his right 'inherit the kingdom'. A wide vocabulary exists to label or insult left-handers, and in many languages the very terms 'left' and 'right' are loaded with bias. From other European languages, and sometimes quite ancient roots, English alone has acquired the terms gauche, sinister and cack-handed, the last being a reference to the custom in predominantly right-handed societies of reserving the left for wiping away faeces or 'caca'. I even have a book on symmetry that contains this entry in the index: 'laevo-, see under: dextro-'. The political left, incidentally, which may or may not have this pejorative connotation according to your own political views, originates in the layout of the French Assembly, where, after 1789, the revolutionaries sat on the left wing.

Left-handers are in a minority, but the true size of the minority is uncertain. In 1942, the psychologist Charlotte Wolff could write without pausing for reflection that 'in these days not more than 2 to 3 per cent of the population are left-handers'. Recent studies suggest, by contrast, that up to a third of children would develop as natural left-handers in the absence of external influences. This matches the apportionment of left- and right-handedness in Palaeolithic man inferred from the way they shaped their axes. But in many environments, there is strong social pressure to be right-handed – even left-handers must shake hands with their right, for example – so the observed proportion of left-handers is often much less than this. United States army recruits, for example, report only 8 per cent of left-handers in their ranks.

The systematic elimination of left-handedness is not what it used to be, though. Listening one day to the statistical curiosities that are the staple fare during boring moments of test match commentary, I was surprised to learn that the first time in cricketing history that the first four men put in to bat were all left-handers was in 2000 – since when it has happened twenty-eight times. On the face of it, this is most odd, given that the statistical record goes back to 1877. It could be down to a chance occurrence of gifted left-handed batsmen, but it's far more likely to be a reflection of the fact that we are now less apt to penalize left-handers in all walks of life. In many sports, of course, being left-handed can give an advantage because all players, even other left-handers, are more accustomed to playing against right-handers.

More insidious pressures remain, however. Almost every one-handed activity, from zipping trousers to using a cash machine, has a cultural bias in favour of right-handers. A visit to Anything Left-Handed, the shop of the Left-Handers Club, which once occupied premises in London's Soho but now trades online, hints at the extent of the iniquity. There are scissors, tin openers, fountain pens and much else in the tiny shop and in its reverse-paginated catalogue. There are also products where many people would never suspect a bias – rulers and tape measures (numbered from right to left), corkscrews (turning anti-clockwise for a better purchase), kitchen knives (serrated on the other side of the blade). The shop also stocks CDs of music played by left-handers, although I'm not sure how you are supposed to hear the difference. But I'm disappointed to find it does not have Ravel's dazzling *Piano Concerto for the Left Hand*. The work was written for Paul Wittgenstein, the brother of the philosopher Ludwig, who lost his right arm in the First World War. When Ravel had finished the work, Wittgenstein proclaimed it too difficult and demanded changes. 'I am an old hand at the piano,' he joked. 'And I am an old hand at orchestration,' Ravel snapped back. Other left-handed pianists hanker for a reversed keyboard with the deep notes on the right allowing the left to carry the melody for once. While we are at the keyboard, we might note that the size of the hands can also affect the kind of music that gets written. Rachmaninov had hands that could span an octave and a half, which puts some of his compositions literally beyond the reach of many daintier players.

The discrimination designed into so many everyday objects may be rather more than an inconvenience. In 1989, the psychologist Stanley Coren surveyed a large number of students at the University of British Columbia and found that a left-hander was almost twice as likely as a right-hander to have a car accident, and half again as likely to have an accident while using a tool of some kind. Coren attributed the cause not to innate clumsiness of southpaws, but to design that is

consciously or unconsciously biased to suit right-handers. He estimated that left-handers' life expectancy was reduced by eight months
in this way.

The hands are not our only handed body parts. Inside the human
trunk, asymmetry is the norm. The heart is on the left and the liver is
on the right. The stomach lies to the left. The left lung has two lobes,
the right has three. There are little-noticed external differences, too.
Our hair falls to one side or another. The left breast is usually a little
larger than the right. The left testicle usually hangs lower than the
right, even though the right one is generally heavier. The reasons for
this aren't firmly established, but it has been known for a long time:
the majority of Classical statues confirm it.

In a way, it is the presence of symmetry in the body at all that is
more remarkable than when it breaks down. The process of embryonic development is one of progressive loss of symmetry. The
fertilized egg is spherically symmetric, but with each cell division it
loses some symmetry. As organisms that must live under the influence of gravity, we quickly lose any top–bottom symmetry. With
locomotion necessarily comes the need to move forward, and so a
sense of front and back, which causes us to lose symmetry in this
direction. That leaves only the symmetry in the third dimension,
from side to side. Here there is no distorting environmental influence, and so the symmetry seen in the embryo can persist. Yet
occasionally, this tidy bilateralism is subverted, and something grows
on one side but not on the other. To see why this is, we must look
more closely at what happens in the developing embryo.

The process of symmetry loss gains pace with the emergence in
the embryo of an arrangement of cells known as the primitive streak.
As growth continues, cells begin to apportion themselves equally
either side of this putative midline of the organism. Although the
same set of genetic instructions is used to make the corresponding
parts on either side of this line, it is something of a mystery as to how
the cells, which seem otherwise identical, direct themselves into mirror-
image positions. The cells may pick up positional information by
detecting variations in waves of cellular activity, rather like a driver

using a satellite navigation system. But this still leaves the problem of directed left–right asymmetry.

An unignorable potential clue to the phenomenon was revealed in 1848, when the young Louis Pasteur discovered that certain chemical molecules exist in left- and right-handed versions. He knew that tartaric acid rotated polarized light (light filtered in a special way) to the right, while tartaric acid made synthetically had no such effect. When he crystallized some of the synthetically made acid, he found he had an equal mixture of crystals that were mirror images of one another. Half were the 'dextrorotatory' form, the one that occurs in nature, and half a new 'laevorotatory' form. It was eventually discovered that many biological molecules, including sugars, amino acids and DNA, have this property. It can matter greatly which form of these substances is present in our bodies, as the author Lewis Carroll seems to have guessed. Lactose and lactic acid are two examples of so-called 'handed' molecules that occur naturally in only one of their handed versions. In *Through the Looking-glass*, Alice holds her kitten up to the mirror and wonders if it would like it there, but then she considers: 'Perhaps Looking-glass milk isn't good to drink.'

It is hard to believe that this molecular handedness is not connected in some way with the overall handedness of biological organisms. So are our left–right asymmetries 'due to some molecular asymmetry that is transferred to a global asymmetry', as the embryologist Lewis Wolpert suggests? If so, how might this scaling-up take place? Wolpert speculates that asymmetric molecules produced along the embryo midline may function chemically to push certain other molecules – and cells – preferentially to one side rather than the other.

A chemical mechanism like this may explain the left- and right-hand biases we (almost) all share, such as having the heart on the left. But what makes nature produce handed molecules in unequal quantities? The answer to this is not known for sure. However, the mirror-image forms of amino acids and other biologically important substances contain one last asymmetry – more of their electrons spin leftwards than rightwards. Might this explain the bias? If so, how did this disparity arise? Perhaps there was some cosmic event to cause it, such as a great burst of polarized light. In that case, there may be another half of the universe where the opposite bias applies.

As for left- and right-handedness in behaviour, the psychologist Chris McManus proposes a genetic mechanism. Handedness may be controlled by two genes, not a *left* gene and *right* gene as you might expect, but one that favours right-handedness, called *dextral*, and one that does not discriminate, called *chance*. Such a mechanism would account for the observed (natural, rather than culturally suppressed) minority of left-handers in the overall population. As is always the case when the talk turns to 'a gene for . . .', it raises the prospect of genetic therapy. We might one day be able to 'cure' left-handedness by suppressing the *chance* gene. But would it not be a sign of our liberation from age-old superstition if instead we chose to suppress the *dextral* gene and leave everything to fifty-fifty chance?

The Sex

The best sight gag in the entire history of art must be the fig leaf. How large it is! And how very suggestive in its shape! How it engrosses what it purports to hide. How many other plant leaves might have done the job with less blatancy. And yet the fig leaf it assuredly is that artists have elected to use when asked to preserve the public decency. The Bible gives them their cover story: in Genesis, when Adam and Eve realized their nakedness, 'they sewed fig leaves together, and made themselves aprons'. But aprons are garments that surely provide rather more coverage than the single, strategically positioned leaf of artistic convention, with its three major lobes simultaneously screening and outlining the penis and testicles behind, and two further vestigial lobes appearing so neatly to represent curls of pubic hair.

The artistic fig leaf became all but mandatory when in 1563 the Roman Catholic Council of Trent ruled that 'all lasciviousness be avoided' in religious images 'in such wise that figures shall not be painted or adorned with a beauty exciting to lust'. Up until that date, in Classical statues and in the Renaissance art inspired by them, false modesty had taken a different form. The human figure was often modelled on the bodies of athletes, who performed their exertions naked. In a monument to a civic dignitary, a philosopher or a general, a well-toned body was the sculptor's means of indicating their good citizenship. The artist would carve the genitals on such statues to be somewhat less than life size. Unless the point was to celebrate fertility, as with the conspicuously erect penis of Priapus, the Greek god so eagerly adopted by the Romans, a sculpted penis of normal size, even a flaccid one, was thought vulgar and a distraction from the achievements that it was the business of the statue to celebrate.

The first nude statue to be put on public display in London since Roman times was the monument raised to the Duke of Wellington not long after his victory over Napoleon's armies at Waterloo. The sculptor Richard Westmacott created a dynamic bronze figure of Achilles so large that it could not be transported into Hyde Park, where it was to be stationed, without knocking down a wall. The artist had taken the precaution of including a fig leaf, but the leaf – and presumably also any genitals that it hid – was comically small. The caricaturist George Cruikshank was not slow to see the potential. His cartoon of the unveiling shows ladies clustered round the statue, which was paid for by a subscription of British women and 'erected in <u>Hide</u> Park', as the caption explains in a barrage of puns. 'My eyes what a size!!' one lady squeals, while another at last locates the salient detail with the aid of a telescope. 'I understand it is intended to represent His Grace after bathing in the Serpentine,' another opines. See, one lady tells the duke himself, 'what we ladies <u>can raise</u> when we wish to put a man in mind of what he has done & we hope <u>will do again</u> when call'd for!' And inevitably, a small child points to the leaf: 'What is <u>that</u> mama'. Immune to such ridicule, the Victorians retrofitted fig leaves to many statues. Even the cast of Michelangelo's *David* in the Victoria and Albert Museum sports this extra adornment.

In the art of the nude, man gains symbolic virtue at the expense of personal identity. After all, Westmacott never would have represented *Wellington* actually naked. Nor would the British public have expected to see their great leader's actual private parts or even his body replicated in bronze. Woman also loses her identity: she becomes simply 'the nude', the generic of female sexuality and vulnerability. The male nude swaggers through the city streets, keeping public decency with his fig leaf. The female nude is for private consumption, preserving her modesty more coyly in what is known in the trade as the pudica pose – with one hand attempting, more or less vaguely, to cover (or is it directing the eye towards?) the genital area. The specialist word in fact acknowledges this ambiguity, stemming as it does from Latin words for both external genitals and shame.

In both cases, we remain uncomfortable about honest depiction of the sex. We have even communicated our prudery into outer space. Michelangelo's *David* may have a small penis, but the gold-plated representation of woman sent out far beyond our solar system aboard the *Pioneer 10* and *Pioneer 11* space probes in 1972 and 1973 has no vagina at all. Why are we withholding the story of our true physical appearance from other species? Will they puzzle over how we reproduce?

The idea that the spacecraft – the first human-made object intended to leave the solar system – should carry some sort of message about the creatures that sent it was enthusiastically taken up by the space scientist and television personality Carl Sagan. At first, only scientific diagrams indicating our location in the universe and one or two other things that we have discovered about it were to be included. But Sagan's artist wife, Linda Salzman, suggested that the graphic should also show a man and a woman. The figures are supposed to have 'panracial' features, to use Sagan's word, though Salzman based them on Greek ideals and the drawings of Leonardo da Vinci. But any fashion-conscious alien would immediately note that their hairstyles alone pinpoint them in the late twentieth century and imply a Caucasian ethnicity. In fact, so provincial do the couple seem, the man waving in greeting, the woman standing demurely at his side, that a satirical magazine in Berkeley ran the image with the caption: 'Hello. We're from Orange County.' As Sagan wrote: 'The man's right hand is raised in what I once read in an anthropology book is a "universal" sign of good will – although any literal universality is of course unlikely.'

The *Pioneer* plaque excited comment from almost every quarter. Women wanted to know why the woman wasn't also waving. Homosexuals demanded to know why homosexual partnership wasn't represented. The art critic Ernst Gombrich pointed out in *Scientific American* that only aliens with a visual system that operates within the same specific region of the spectrum as our own would be able to see the image.

But the most heated debate swirled around the nakedness of the humans and their visible or invisible sex organs. The two figures stand slightly apart rather than holding hands as initially conceived, so that they are not misconstrued as a single hermaphroditic organism. But other than this subtlest of clues, there is little to indicate that

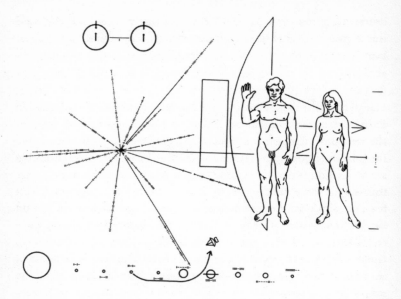

we are a species reliant upon sexual reproduction, which seems a significant omission considering that this phenomenon remains one of
the deepest oddities about life on earth. Reprinted in newspapers, the
design drew predictable accusations of pornography. The *Philadelphia
Inquirer* took the precaution of erasing the woman's nipples and the
man's genitals. The *Chicago Sun Times* progressively amended the
versions they printed in successive editions of the day's paper to
obliterate the man's genitals, too. On the other hand, the incompleteness of the woman's drawing also provoked complaints of censorship.
Sagan defended the omission of a line representing the vagina on
grounds of artistic tradition, although it seems that he and his wife
took the decision at least partly to head off any difficulties with puri-
tans among the NASA top brass.

Sagan pointed to ancient Greek statuary in particular, although
the Greeks created few statues of women other than Aphrodite, the
goddess of beauty. For the most part, during both the Classical and
neoclassical periods, artists preferred to duck the whole issue by the
use of the pudica pose or a strategically draped cloth, but it is true
that female nudes made without these devices did, like the *Pioneer*

drawings, generally omit any hint of a vagina. As Sagan himself noted, the real value of the whole episode was to raise the issue of how we represent ourselves to ourselves more than to any other species.

The French twentieth-century philosopher Roland Barthes made much the same complaint of incompleteness when discussing Parisian striptease in his *Mythologies*. 'Woman is desexualized at the very moment when she is stripped naked,' he discovers. (We imagine the poor ecdysiast doing her best as the learned semiotician sits in his pullover and tweed jacket, taking notes.) 'We may therefore say that we are dealing in a sense with a spectacle based on fear, or rather on the pretence of fear, as if eroticism here went no further than a sort of delicious terror, whose ritual signs have only to be announced to evoke at once the idea of sex and its conjuration.' The fig leaf presents a vegetable barrier to the fleshy, animal sex. The diamanté g-string that is revealed at the (anti)climax of the striptease, Barthes groans, presents an impenetrable mineral barrier. 'This ultimate triangle, by its pure and geometrical shape, by its hard and shiny material, bars the way to the sexual parts like a sword of purity.'

We can't leave it here. John Donne goes all the way in his elegy, 'To His Mistress Going to Bed'. All such obstacles are removed one by one in this poet's striptease: 'Unpin that spangled breast-plate, which you wear, / That th'eyes of busy fools may be stopp'd there.' Then, 'busk' and gown and hose come off, until at last . . .

> O, my America, my new found land,
> My kingdom, safeliest when with one man mann'd,
> My mine of precious stones, my empery;
> How blest am I in this discovering thee!

I am perched on the edge of April Ashley's bed at her flat in west London, flipping through her personal memorabilia in a cardboard box. There are copious newspaper and magazine cuttings – 'My Strange Life, by April Ashley', 'The Sailor Who Made a Fair Lady', 'Sex Op Girl Weds Again' – as well as modelling photographs and a California car licence plate that reads 'APRIL'. I'm looking for the identity documents that April has assured me with an imperious wave of her hand must be in there somewhere.

April Ashley was one of the first people in Britain to undergo a full sex-change operation. (This procedure is now generally described as gender reassignment, which is both more sensitive and more biologically accurate, as we shall see.) She was born a boy – George Jamieson – in Liverpool and grew up there in a large family during the Second World War. 'Although I was brought up a strict Roman Catholic boy, I knew from the age dot that I was a girl,' she wrote later. George joined the merchant navy at the age of fifteen, making his way via a succession of jobs to London and then to Paris and the stage of the Carrousel night club, famous for its female impersonators, where she adopted the stage name Toni April. She began female hormone treatment in order to accentuate her femininity, but believed only surgery would bring about the full alignment between the sex she felt herself to be and the sex she by now appeared to be that would allow her to go on living. In May 1960, aged twenty-five, she travelled to Morocco and had a surgical operation to remove the male genitals that she felt were not hers and to construct a vagina in their place. Returning to Britain, she was required to change her name by deed poll and, as April Ashley, began a long struggle for official recognition as a woman.

I find the documents I am looking for – cancelled passports, a marriage certificate, a United States resident alien card, and a birth certificate reissued in 2006. There are many ways to tell a person's story – or, to put it more honestly, to make a story of a person. Photographs and official documents are just the most obvious and conventional way to do it – the one accepted by authority. It occurs to me that you could tell April's story very well in shoes – the wooden clogs in increments of size that George wore growing up in the slums of Liverpool, the deck shoes of a merchant seaman, sexy heels in Paris, and the more sensible shoes of mature womanhood. That at least would have character. As it is, the official scraps of paper that mark the progress of our lives often seem to miss out what really matters to us. In April's case, they are hardly up to the task.

April – George – was male at birth, as the paperwork records. The fact that he did not feel himself to be male as he grew up is nowhere to be found. As we saw when discussing the face, society requires us to actually be what we appear to be, with little regard for all the other things that we might feel ourselves to be. If you have male genitals, you tick the M box on the form. If you have a vagina, tick F. They are the only options. So far as officialdom is concerned, sex and gender are one and the same. Only after she had had her operation was April able to change her name and obtain a passport in her female identity.

April has been married twice. The first marriage was not a success, and her first husband filed for an annulment on the basis that April had been of the male sex at the time of the marriage, even though the marriage took place after she had had her operation, with a new passport used as proof of identity at the ceremony. The case came to court in November 1969. April underwent physical and psychological examinations by medical teams for both the prosecution and the defence. They showed her to have normal male XY chromosomes, but she scored towards the 'female' end of the sexual spectrum in a psychological test. In a controversial ruling with far-reaching implications, the judge disregarded April's psychological profile and the fact of her surgical alteration, and declared that the 'true sex of the respondent' was that indicated by the chromosomal evidence and original anatomy. The case provided a legal precedent for deeming a person's sex in English law to be that which it was at birth, regardless of their subsequent gender history. Only in 2004 was the law liberalized to allow transsexual people to be recognized in the gender to which they have transitioned. The Gender Recognition Act now provides for the amendment of birth certificates to show the new gender. This enables people who have undergone gender reassignment to keep their former gender confidential from employers and partners.

There are many ways in which biological sex may not correspond with psychological gender, some of which challenge social norms and cause consternation in law. At the fundamental level, there are chromosomal variations. At conception, it may surprise some to learn, we are all essentially female. Although the woman's egg contributes an X chromosome and the man's sperm either an X or a Y chromosome, these do not immediately determine the sex of the embryo. At eight weeks' gestation, the fertilized egg is implanted in the uterus. If it has a Y chromosome, it then responds to a chemical signal that causes testes to begin to form, and the potential female reproductive system to wither. If not, it continues in its 'default setting' until, at thirteen weeks, the foetus gonads begin to transform into ovaries.

In a small proportion of people, the chromosomes do not pair up properly. An extra chromosome may cause a male to be born XYY, a so-called 'super male', or XXY, with low testosterone and a low sex drive. Such people usually look male and think of themselves as male, although they may have small genitals and the beginnings of breasts. Barry (later Carolyn) Cossey was born not XXY, but XXXY, with two extra chromosomes. He later underwent an operation for gender reassignment and appeared briefly as a Bond girl in the film *For Your Eyes Only*. A female may also be born XXX. An XO person, in which the second sex chromosome never appears, on the other hand, may have female-looking genitals but no ovaries. In addition, environmental stresses on the mother during the early weeks of pregnancy can alter the balance of hormones released in the womb, causing physiological changes in the unborn child. These variations can lead to a wide range of chromosomal, gonad, genital and hormonal anomalies. Overall, so-called intersex conditions of one sort or another may affect as much as 2 per cent of the population. One consequence of April's operation was to make it impossible to ascertain whether she was born intersex.

True intersexuality, where a person has clear sexual characteristics of both sexes, such as an ovary on one side of the body and a testicle on the other, is extremely rare. The traditional catchall term for such conditions is hermaphrodite, from the name given to the offspring of the Greek gods Hermes and Aphrodite. Yet the original Hermaphroditos was not born intersex. In Ovid's *Metamorphoses*, Hermaphroditos is a beautiful boy who bathes in the pool of Salmacis. There, a water nymph wraps herself around him, and their bodies are merged into one, 'which couldn't fairly be described as male or female. They seemed to be neither and both.' The simplest explanation for the story may be the disappointing effect of the cool water on the boy's anatomy. Emerging from the water (perhaps like Wellington from the Serpentine), he sees that 'the pool which his manhood had entered had left him only half a man'.

This is one of several sexual transformations in *Metamorphoses*. Another story concerns Iphis, a girl who has been brought up as a boy because her father has warned her mother that she must kill any girl child she bears. The day comes when Iphis is due to marry, and after a desperate appeal to the gods, she leaves the temple miraculously transformed into a man, 'with longer strides than she normally took', a darker complexion, more angular features, and even cropped hair. In another tale, beautiful Caenis is granted her wish to be transformed into a man by Neptune in compensation for his having raped her. Pleased with the effect, Caenis, now Caeneus, thereafter 'devoted his life to manly pursuits'.

These ancient stories are a reminder that our sex and our sexual identity have not always been regarded as fixed entities. Before chromosomes were understood, and when examination of internal sex organs was still impossible, the line between biology and psychology was less neatly drawn. It is an irony that the modern possibility of surgical transformation may in fact be reinforcing a social view that these things should stay as they were found to be at birth, or at least firmly where they are subsequently put. Psychologists, meanwhile, often speak of a sexual spectrum. The spectrum view is helpful because it suggests the possibility of intermediate positions along the way, as well as concentrations at opposite 'male' and 'female' ends. But it may not be quite the right analogy, suggesting as it does that, as you slide towards one end of the scale, you necessarily slide away from the other.

For, biologically speaking, sex is not a zero sum game. The 'male' hormone testosterone and the 'female' hormones oestrogen and progesterone are all present in both men and women. They perform a variety of functions in addition to their well-known roles in sexual development. Levels of these hormones typically reinforce the apparent sex of the person, with an average of fifty times more testosterone in men than in women, for example. But the *range* of concentrations actually overlaps in men and women, so that some men have less testosterone than some women, and some women have less oestrogen or progesterone than some men. Nevertheless, the popular image that there is one chemical essence for men and another for women is hard to dislodge, and we are likely to be stuck with 'testosterone-driven' footballers and stock market traders for a long time yet. Strangely, women are never described as 'oestrogen-driven', although they may occasionally find themselves labelled broody or mumsy.

Experiments in which animals are treated with these hormones now show that, contrary to previous belief, 'maleness' and 'femaleness' are independent variables. Females of various animal species given testosterone, for example, begin to exhibit typical male behaviour, including attempting to mount other females, but this is not accompanied by a loss of female behaviour. What this seems to indicate in humans is that gay men, for example, may be somewhat female-like, but simultaneously as 'male' as straight men. In general, homosexual people may be sexually more like people of the sex they do not belong to, but no less like their own sex than heterosexuals. Bisexual people may be not simply interested in sex with both men and women (or confused, as some heterosexuals would have it), but just more interested in sex, perhaps because they received a larger dose of prenatal hormones. Both aggressively heterosexual and campaigning gay neuroscientists have tried to locate regions in the brain that would 'explain' gayness. But these behaviours don't require exceptional explanation so much as fitting into a picture that encompasses the full range of permutations of biological sex, psychological sexuality and sexual preference.

To this natural biological variability, we must add cultural factors. Gender refers to our social and cultural self-definition as distinct from our biological sex. Our expectation of what gender is and ought to be is shaped by culture, and one of the principal restrictions is the existence of gender in grammar. Why, in French, is a table feminine and a desk masculine? Why, for that matter, is a table feminine in French but masculine in German? The nonsense of all this is revealed in the fact that words for the sex organs themselves frequently have the wrong gender. In French, for example, 'la bite' is the cock, and 'le con' is the corresponding slang term for the female genitals (and less offensive than its English equivalent). In Greek, Marina Warner observes, the words for knife, fork and spoon have three different genders. Gender is a superfluous development in grammar that should gradually wither from the world's languages, say the experts, although perhaps it will not do so quickly: in long ungendered English, ships are still designated feminine (even the USS *Benjamin Franklin* and HMS *Nelson*). The word 'gender' merely means 'type', and has no intrinsic sexual significance. Where there were two (or three) such types, it simply happened that grammarians chose to call those types masculine and feminine (and neuter). They might as well have called them left and right or up and down or black and white.

Gender is one of the major ways in which we continually reinvent ourselves. Lifelong, we respond to those around us by performing to our chosen gender identity in order to meet – or occasionally to challenge – social expectations. The clearest example of this is perhaps the contemporary pressure to dress baby boys in blue and baby girls in pink. This is a cultural consensus with no basis in biology. In Victorian times, by complete contrast, children were often dressed alike in smocks until the boys were 'breeched' (put into trousers) around the age of six. Passports and public lavatories oblige us to choose one of two genders. Even language can force us to make a declaration, as in some languages the same phrase will have different word endings whether it is uttered by a man or a woman. In our bodies and minds, though, we may feel only *comparatively* male or female, rather than totally and unambiguously gendered in this way. Moreover, this strength of genderedness may alter during our lives.

There are numerous examples in fact and fiction of both men and women who have spent an extended period of their lives presenting themselves as other than their birth sex, including the legend of the female Pope Joan, who supposedly held office in the ninth century, although this is probably a later fabrication designed to discredit the papacy. Here are two stories from the eighteenth century.

The French diplomat and spy Chevalier d'Eon de Beaumont claimed to have been born a girl in 1728. He was brought up as a boy, possibly in order that his parents could gain an inheritance that was conditional on their producing a male heir. He became a spy for Louis XV and fought in the Seven Years' War, but eventually came into disfavour and was pensioned off to live in exile in London. Here, his feminine looks gave rise to gossip, and a gambling pool was started on the London Stock Exchange as to his true sex – the bet was never called, however. After the death of Louis XV, d'Eon petitioned to return to France as the woman he now claimed to be. His wish was granted provided he dress as a woman. Horace Walpole met d'Eon, and later noted: 'her hands and arms seem not to have participated in the change of sexes, but are fitter to carry a chair than a fan.' A post-mortem found that she had been a man all along.

Hannah Snell, born in Worcester five years earlier than d'Eon, travelled in the opposite direction. When her marriage broke up after the death of her child, she adopted the identity of her brother-in-law and joined the Royal Marines in pursuit of the husband who had deserted her. She had enjoyed playing with toy soldiers as a little girl. Now, she took part in the British campaign in India. She was wounded eleven times, including once in the groin. It is assumed she must have treated the wound herself or else relied on the help of a sympathetic Indian nurse for the truth not to emerge then. In 1750, her ship returned to England, and she revealed her true sex, capitalizing on the ensuing sensation by selling her story to the press and giving performances on the stage. In later years, she ran a pub in Wapping which she named 'The Widow in Masquerade, or the female Warrior'.

Before it was relatively easy to undergo gender reassignment surgery, and before science possessed the wherewithal to think of locating spots in the brain that might determine what we think we are, the experience of shifting sexuality was less a problem to be solved and more part of life to be lived. It is an ironic consequence of the possibility of gender reassignment that our cultural ideas about sexual identity have become more fixed, rather than less.

The Foot

After fifteen years alone on the island where he has been shipwrecked, Robinson Crusoe one day spots a single footprint on the sandy shore. Left or right, large or small, he does not say. Nor does he immediately do the obvious thing and place his foot alongside the print to confirm that it is not actually he who has made the mark previously.

The footprint makes its appearance exactly halfway through Daniel Defoe's famous story. From the moment he is washed up on the shore to this point, however, there have been repeated hints that Crusoe is not completely alone. He fears cannibals even though he believes the island to be deserted. He sees a vision of a man calling on him to repent his sins. Some creature tramples his food. He even hears speech – but it is only his parrot, Poll.

The footprint is the first real evidence of another living human presence. Three days after he first sees it, Crusoe at last considers the possibility that it could be his, and rules it out by measuring it against his own foot, which turns out to be 'not so large by a great deal'.

Crusoe eventually learns that the island is occasionally visited by cannibals who bring their victims there for slaughter. When a suitable opportunity presents itself, he puts into practice a dream he has had of saving one of these prisoners from the pot. The rescued 'Indian', whom he names Friday, becomes his 'servant . . . companion . . . assistant'. Now, what about the footprint? It is quite clear that the footprint is most unlikely to belong to Friday, despite the popular belief that it does (a belief apparently shared by Umberto Eco, who understands the footprint in this way in a discussion of signs and clues in his *Theory of Semiotics*). Seen some time earlier, it's obviously much more likely to belong to one of the cannibals or their captives from a previous visit to the island, although in fact we never learn whose print it is. It remains simply a clue, a generic sign of human presence.

This is not to say that the footprint is without further meaning. A footprint is many things. It is, for instance, a claim to possession of land. Such a sign is often swiftly followed by the still more assertive action of planting a flag, as Neil Armstrong's boot mark in the dust of the moon reminds us. On Crusoe's island, tellingly, it is (presumably) an indigenous savage who has left the footprint, but it is Crusoe who claims 'undoubted right of dominion' over 'the whole country'.

In its isolation, though, this footprint is more intensely symbolic than just this. A trail of footprints would suggest a particular person, a body with direction and purpose, the path of a hunter perhaps. But a single footprint in the sand raises the question of how it got there. In this sense, it is a divine symbol, an indication that Crusoe need lack neither god nor human fellowship. For the gods and holy men leave footprints – Christ does so on the Mount of Olives, and Muhammad does at Mecca, while both the Buddha and Vishnu measure out the size of the universe with their steps. Such contact with the ground clearly demonstrates earthly concerns.

In *Enquiries Concerning Human Understanding*, David Hume uses the hypothetical circumstance of finding just such a lone footprint to consider the question of whether there exists 'a Particular Providence' or God. 'The print of a foot in the sand can only prove, when considered alone, that there was some figure adapted to it, by which it was produced: but the print of a human foot proves likewise, from our other experience, that there was probably another foot, which also left its impression, though effaced by time or other accidents,' he writes. On the other hand, Hume reasons, 'The Deity is known to us only by his productions.' Yet from these productions, the wonders of nature, we cannot infer anything directly about Him because, unlike in the case of the footprint, we have no other knowledge to adduce. We are human, and know the shape of a foot and the mark it leaves, but God's productions – if that's what they are – lack this reference. Nature's works cannot therefore be taken as proof of His existence. And more than that: 'All the philosophy, therefore, in the world, and all the religion, which is nothing but a species of philosophy, will never be able to carry us beyond the usual course of experience, or give us measures of conduct and behaviour different from those which are furnished by reflections on common life.'

The Czech author Karel Čapek – whose play *R. U. R.* (Rossum's Universal Robots) gave us the word 'robot', and of whom more later – makes this logic his point of departure in a humorous short story called 'Footprints'. Mr Rybka is walking home in the fresh snow, speculating idly on the owners of the various prints he sees. Then he notices footsteps directed towards his own house. 'There were five of them, and right in the middle of the street they came to an end with the sharp impression of a left foot.' Unsettled now, Rybka opens his door and calls the police. The sergeant turns up, deduces various things from the prints – hand-sown shoes, brisk stride – and assures Rybka that, since no deeper impression was left in the toe of the final print, the person who left them cannot have simply jumped off somewhere. Where did he go then? Why do the prints just stop? The sergeant can go no further – no crime has been committed. But a man has disappeared, Rybka insists indignantly. The police, the sergeant chides him finally, are interested in misdemeanours, not in mysteries.

What can we tell from a human footprint? Certainly not whether its owner is savage or civilized. The moral of *Robinson Crusoe* revolves around the question of who is the more civilized, Crusoe or islander, with the Englishman forced to consider that right is not all on his side. We can, though, perhaps infer the nature of the master–servant relationship that Crusoe imposes. Friday's feet are larger than Crusoe's, but in a bizarre scene Friday kneels before Crusoe, lays his head on the ground and then places Crusoe's foot upon it, a gesture that Crusoe interprets as a 'token of swearing to be my slave for ever'. A few pages later, Crusoe's reading of the symbolism is made plain when he teaches Friday that 'God is above the Devil, and therefore we pray to God to tread him under our foot'.

Fossil footprints enable scientists to glean rather more information about those who made them thousands or even millions of years ago. The shape of a foot reveals which of various hominid species may have passed that way. Height may be estimated from the size of prints using a conversion factor based on anthropometric data. Speed of walking or running can then be deduced from the length of the stride. The depth of a foot's depression shows where the greatest pressure has been applied, and this information can be used to draw conclusions about gait. Was the walker creeping up on prey? Was she carrying a child at her hip? Or was he carrying a dead animal across his shoulders? Prints may even be dated with some certainty by analysing the way different ingredients in the soil have been rammed together when the impression was made.

In 2005, Australian anthropologists reported the discovery of fossil footprints from the Pleistocene era about 20,000 years ago in the Willandra Lakes region of New South Wales. The marks of a number of adults and children were preserved. The tracks of one man, designated T8, showed that he had been running in the thin layer of mud fringing the lake. From the placement and depth of the depressions and the distance between them, the scientists estimated the man's speed as a respectable twenty kilometres per hour. A year later, however, Steve Webb, the lead scientist reporting these findings, re-examined the site after new tracks had been uncovered, including four new footprints of T8 (bringing his personal tally to eleven among hundreds of individual prints). This time, he came up with a very different estimate of T8's speed: thirty-seven kilometres per hour – faster than Usain Bolt, the present world-record-holding sprinter, would be able to run on the same surface. The discovery caused quite a stir, and provided useful ammunition for Peter McAllister in his book *Manthropology*, which catalogues the supposed physical inadequacies of modern man. In addition, Webb declared incredibly, T4 was the trail of a one-legged man moving at a clip of 21.7 kilometres per hour, leaving the prints of one foot and the post marks of his crutch. This unlikely revelation was prompted by consultation on the interpretation of the tracks with Pintubi people of Central Australia who still hunt by tracking on foot. Their reminiscences about one of their number who had lost a leg but nevertheless remained highly mobile in the field encouraged Webb to draw this bold conclusion.

Another tantalizing trail was set around the same time when another cluster of prints was found in volcanic ash on a dried lakebed in central Mexico. Tracks showed birds, livestock and domestic pets as well as adults and children, possibly all fleeing together from a volcanic eruption. The depressions in the compressed ash were initially dated as being up to 38,000 years old. Since it is thought that humans first came to the Americas less than 15,000 years ago, the discovery was set to revolutionize human archaeology. Either the North American continent had a human population far earlier than had been previously thought or there was a gross error in the dating of the prints. A second team of scientists then dated the ash itself to 1.3 million years old, well before the appearance of humans anywhere on the planet, forcing the first team to look again at its data and to concede their error. The prints are damaged where water has flowed across them so that even a clear left–right pattern in them is hard to discern. Could they be the prints of some much earlier hominid? Or are they a complicated mixture of modern footmarks and other manmade traces in ancient ash – much as that last footprint in the snow in Capek's story becomes when, on the final page, the police sergeant's colleague arrives and inadvertently treads on it with his boots? It seems that, as Mr Rybka finds, reading footprints of whatever age is an unreliable art.

A footprint, then, is not only the mark of a person who has passed that way, but also a relic of dynamic human action in the past. That person long ago walked, or ran, or crept and sprang after prey, or fled from danger. The foot is a site of extraordinary power, not only the launching pad of physical action, but a part of the body identified in older belief with generative potential. Three thousand years ago, it was believed that the first Chinese emperor of the Chou dynasty was born as the simple consequence of his mother's stepping into the footprint of the deity. Well into the modern period in China, it was the tradition that husband and wife were not allowed to see one another's feet because of their procreative significance. So strong was this taboo that women's feet were bound away from sight so tightly that they were often permanently deformed. This bashfulness has its echo in the West, occasioning the famous myth that the Victorians even swathed the feet of their pianos, though it seems it truly was a myth: Victorian catalogues openly advertised pianos with naked legs, for one thing, and 'even in Victorian times it surfaced as a satirical joke', according to Ruth Barcan's study *Nudity*.

The human body that we know best is not the anatomist's cadaver or the sculptor's stony ideal but our own, in constant life and motion. Its most animated actions involve varieties of bipedal motion. Today, sport provides a ritualized vestige of these actions required for fight, flight and survival. The pentathlon of the ancient Olympic Games involved five such demonstrations of agility that still feature in sporting events today: a running race, long jump, javelin and discus throwing, and wrestling. The addition of such cultural artefacts as balls, measured fields of play and formal rules gradually distanced us from these primitive activities, and gave the feet more exacting athletic tasks to perform, such as kicking to score a goal.

The form of movement that more greatly interests me, however, is dance. Here, extreme physical exertion must be combined with extreme restraint to produce artistic expression. It is at once a highly sophisticated and yet also strangely primal activity. If sport is our cultural legacy of the actions necessary for individual survival, then dance, it seems to me, is our inheritance from our first attempts to make connections. It contains the erotic, the religious and, in the synchrony of a war dance or of a *corps de ballet*, the urge to be one of the group. Dance is the bodily expression of civilization.

I have come to learn more from Deborah Bull, a former principal dancer in the Royal Ballet at Covent Garden in London. I saw her in several roles in her heyday. The performance I remember best was in an inventive ballet describing the plight of endangered species, set to the music of an ensemble called the Penguin Café Orchestra. Deborah was a Utah longhorn ram, a role that demanded a lot of mournful jerking around the stage while wearing a cumbersome headdress, and a complete surrender of the feminine elegance one usually associates with the ballet. Today, she sits smartly dressed in cream and black in a windowless office at the Royal Opera House, where she is the creative director. A poster for the 1948 London Olympic Games hangs on the wall. She jiggles a sandal negligently on her otherwise bare foot as if to remind me of my purpose.

The rules of ballet, Deborah tells me, developed at the court of Louis XIV. These rules may seem arbitrary now, even capricious, but they arose out of the fashions and customs of that time. They decree that particular physical actions must have a particular look. 'In sport, it doesn't matter what you look like. A footballer can score a goal any way. But a ballet dancer must move their leg *in the right way*.' For example, the stance known in ballet as 'turnout' – standing with the heels together and the feet pointing outward in a straight line – may have developed because the king, himself a dancer, would turn out his feet so that people would admire his silken shoes. The action appears highly stylized to us now. In fact, it looks impossible. But as I find to my surprise, even I can stand in this way without too much difficulty. Standing like this makes me newly aware of the working of the major muscles and joints in my legs. I feel an unaccustomed strain in the ligaments of my hips, for example, that a trained dancer with more elastic ligaments would not feel. More importantly, I realize that my proprioception – my sense of my own body's position in relation to itself and its surroundings – is being awakened.

Standing *en pointe* – with the body's entire weight borne on the front of the toes – is something I do not attempt. The position developed as a way for dancers to appear elegantly lighter than air, as if hovering a few inches above the ground, and seems still more unnatural. I say so, and discover that Deborah has a bit of a bugbear about the supposedly torturous demands of ballet. 'Ballet develops the muscles to hold the skeleton in a particular shape,' she tells me a little severely. 'And developing a muscle is not a bad thing to do.' *En pointe*, the foot makes the terminus of a straight line to the ground, which is braced all the way up the body by the muscles of the calf, thigh, abdomen and back in turn. I am reminded of the structural engineer's view of the body as a system of columns and beams and levers. I see that, held like this, the body's entire weight is constantly brought back towards this central axis, so that it passes down the leg and out through the pointed foot. It is like the steel column of a modern building, which tapers almost to a point where it touches the ground despite the great weight that it supports. 'These movements are totally within the bounds of the human body,' says Deborah. 'We don't know its limits.'

The concept of record-breaking so crucial in sport is absent from dance, but there is nevertheless a drive for constant physical improvement. So, for example, in the arabesque – a movement where the dancer stands on one foot and extends the other behind her – the height to which the dancer raises that leg has crept up over the decades. There are some fundamental limits, however: the height to which a dancer can jump has remained constant because it is subject to immutable physical laws. (In fact, to a good approximation, it is the case not only that all comparably fit humans, but also that all species capable of jumping, from the flea to the elephant, can jump to roughly the same absolute height of a metre or so. This is because both the energy needed to produce the jump, generated by the muscles, and the potential energy gained at the top of the jump are directly proportional to the animal's mass, ultimately making this mass, or size, an irrelevant consideration.)

Above all, the physical activity of dance is distinguished from sport by the requirement to disguise the effort involved. Watching a sport, we hear the grunt of the wrestler, see the sweat of the treadmill runner and notice the wobble as a weightlifter's leg threatens to give way under him. Some of these signs of exertion may be solely cultural, that is to say physically avoidable, and only used by the sportsperson as a way to show how hard they are trying. It's hard to believe that the extravagant shrieks with which some tennis players now embellish their stroke-play are anything other than part of a performer's act, for example.

But grunting is clearly out in ballet. So are visible sweating and wobbling limbs. All of these would shatter the outer shell of effortlessness that the dancer must project in order to produce art. At the Laban Dance Centre in its colourful modern home on Deptford Creek in south-east London, I learn of a scientific project to explore the physical limits of the dancer's body where this illusion breaks down. *In Preparation* is a twenty-minute dance piece that aims, according to its creators, to expose 'the effort beneath "effortless"'. The performer and subject of the experiment will be Emma Redding, a dance scientist at the centre. The choreography will require her to execute repeated strenuous actions until she is overcome by muscle fatigue and exhaustion. The 'performance' will be based not on what she thinks her limit is, because we naturally tend to stop ourselves before we reach it, but the further limit to which she is pushed by a trainer. 'Near collapse,' Emma tells me, 'someone would be getting nauseous and light-headed and be shaking. Which are our anthropological habits and which things do we need biologically?' Emma's legs will be strapped with devices to monitor the build-up of lactate in her muscles and other vital signs. The scientific data will be interpreted along with more subjective feedback, such as Emma's running commentary on what she is feeling and the critical comments of observers.

The latter are valuable because the impressions of knowledgeable spectators of physical activity are especially informed. They are known to be based on the action of so-called mirror neurons. Discovered as recently as 1992 during the course of magnetic resonance imaging studies, mirror neurons are brain cells that fire not only when you perform a particular action for which you have been trained, but also when you see that action performed. The phenomenon helps to explain why the best sports commentators are often people who have themselves performed at a top level in that sport. When the football commentator sees the footballer's foot flick the ball in a certain way he is able to predict the flight of that ball more accurately than a naive spectator. It also explains why dance critics are more often ex-dancers than music or theatre critics are former musicians and actors. They quite literally have a feel for what they are seeing, and use this extra-sensory dimension to inform their judgement. On a broader level, it seems that mirror neurons may play a vital role in helping us to learn through observation, and may also be involved in our ability to feel empathy.

Emma seems amazingly upbeat about the prospect of what might in other circumstances be fairly described as torture. No dancer myself, I cannot engage my mirror neurons properly to empathize with her. All I can do as I leave is wish her well with the experiment.

★

Perhaps because they are usually in motion of one kind or another, feet when they are frozen in stone seem to exert strange powers. In the Book of Daniel, the Babylonian king Nebuchadnezzar has a terrifying dream in which an idol appears before him with a head of gold, arms of silver, a trunk of brass, and 'legs of iron, his feet part of iron and part of clay'. It seems that the further the parts of the figure are from the ground, the more precious and artificial they become. The feet of clay – a metaphor for the fragile unity of his kingdom and one that is still used today to indicate a person of flawed character – are helplessly rooted to the earth. It is perhaps a warning to the king that he should not neglect the soil over which he rules.

Another 'king of kings' is the subject of Percy Shelley's familiar 'Ozymandias', inspired by the remains of the vast tomb of Rameses II, the thirteenth-century BCE ruler of Egypt, who reigned some 700 years before Nebuchadnezzar. The poem is another dream-like vision, and indeed Shelley worked from an ancient Greek historian's account of the monument, which was already a ruin in the first century BCE and long gone altogether by the time that he was penning his famous lines in 1817. Its anonymous narrator reports the description by a 'traveller from an antique land' of the ruined statue at Thebes, of which only 'Two vast and trunkless legs of stone' remain. The poem was written in a contest with his friend Horace Smith, whose own effort reduces the remnant to a single limb:

> In Egypt's sandy silence, all alone,
> Stands a gigantic Leg, which far off throws
> The only shadow that the Desert knows: –
> 'I am great OZYMANDIAS,' saith the stone,
> 'The king of Kings; this mighty City shows
> 'The wonders of my hand.' – The City's gone, –
> Nought but the Leg remaining to disclose
> The site of this forgotten Babylon.

To gain an impression of the residual power of even just a foot, I recommend a visit to the Capitoline Museum in Rome. Here are the monumental surviving fragments of the memorial to another great ruler, the Roman emperor Constantine. The so-called *Colossus of*

Constantine once stood some twelve metres tall in a basilica at the Forum. But today all that remains are the head, the right arm, two right hands (it has been suggested that the statue was remodelled at some point so that the emperor was seen holding a Christian symbol), both kneecaps, some fragments of shin and the feet, which are so large that you need to use both arms to encircle one big toe. The reason why it is the extremities alone that survive is that they were carved in marble rather than assembled from soft brick like the bulk of the statue. When the fragments were uncovered in 1487, history reminded us of the parts that most make us human.

The Skin

Perhaps as early as the fifteenth century, a variety of rose arrived in France from its native Crimea to be named Cuisse de Nymphe, which means 'thigh of nymph'. The flower was an extremely pale pink with a slight tinge of lilac. In 1835, the winemaker Laurent-Perrier gave the same name to a new rosé champagne. In Britain, Victorian prudery saw to it that the rose variety was renamed more modestly as Great Maiden's Blush. (It produces a large, full blossom; the English name is not a reference to ladies' girth.) No such strictures applied in France, though, so when a new sport of the rose was developed with a deeper pink flower, it was named Cuisse de Nymphe Emue – thigh of *aroused* nymph. This rose was a favourite of the writer Colette, and makes a brief appearance in her semi-autobiographical novel *Sido*. The colour has been dismally translated into English as 'hot pink'. This shade, too, soon spread to other things. Amid the rainbow of synthetic colours that became available for artists' paints during the mid nineteenth century, often given the names of recent European battles such as magenta and solferino, was also a cuisse de nymphe émue, although its precise hue was apparently rather variable, 'anything from pink to lilac to yellow'.

Painting flesh has always been one of art's greatest challenges. It is not a colour that comes ready-mixed out of any tube, not least because individual skin tones vary so much. It emerges instead from the skilful mixture of the four basic colours of the ancient palette favoured by artists such as the Greek Apelles: red, yellow, black and white. These colours were associated with the four elements, and by extension therefore also with the four humours. Mixed in different proportions, they could represent any shade of skin, from the palest infant to the tanned sailor, and from the florid bacchanalian reveller to the most deathly cadaver.

Realistic skin tone – or rather, realistic variableness of tone across a person's skin — is something scrupulously avoided in most three-dimensional representations of the human body. Barbie's skin exhibits no flaw (unless you can find the umbilicus where she was injection-moulded), but also no variation in colour due to veins and blood vessels. She has no scrofulous patches, no body hair, not even tan lines. She is supremely glabrous. Shop-window mannequins like-wise may have preternaturally pert nipples but they never seem to have areolae darker than the rest of their plastic skin, as we do.

When the natural variation and texture of skin is faithfully repro-duced, as in the sculptures of Ron Mueck, the effect can be unsettling. The son of toymakers, Mueck began his career making animation models for Australian television and the advertising industry, only latterly becoming an artist. A work of 1997 called *Dead Dad* gives a good idea of his technique. It is a supine figure of the artist's deceased father, not much more than one metre in length, so about half life size. It reproduces his skin, which is pale with a slight shine, edging to pinkness at the ears and eyelids. Each crease on the knuckles is there, and every piece of stubble on the chin. The work is disturbing not only because of its highly personal nature, but because it has Mueck's characteristic combination of extreme realism at the level of detail and the gross inaccuracy of its scale. It brings our perception and our experience into direct conflict, asking us, forcefully, to believe that it is real at the same time as it tells us, equally forcefully, that it is not.

What all these bodies lack, of course, apart from a third dimension or the correct magnification, is life. Barbie's perfect skin is repellent to the touch because it is hard, cold and sticky-slippery, whereas we know that our real skin is warm, either soft or firm, and delightful to caress. The warmth comes from circulating blood, which also pro-vides the surface colour that distinguishes the live body from the corpse. A human being radiates power at a rate of 100 watts when resting, rising to 300 watts when doing exercise, which is a power conversion per unit area roughly equal to a rooftop photovoltaic solar panel, and enough that architects must take it into account when designing spaces that will be crowded with people. This heat is usually a welcome sign of life. We prefer a warm handshake to the one proffered by a 'cold fish'. But sometimes it is a reminder of human presence we would rather not have. Marcel Grossmann, a Swiss mathematician who was a student contemporary of Einstein, once confided to the physicist that he could not happily sit on a warm lavatory seat, to which Einstein blandly pointed out that the heat is 'entirely impersonal so that to receive it in this way was not to be subject to an unwanted intimacy'.

I do not know if Charles Darwin grew Cuisse de Nymphe roses (émue or otherwise) in the garden where he took his daily circular walks, but he did concern himself with the question of maidens' blushes. In fact, it was something that bothered him for most of his working life. He made his first notes on the topic in 1838, speculating that dark-skinned people surely blush just as Europeans do, and that animals do not – he was almost sure he had seen a Tierra del Fuego woman blush when he visited that land during the voyage of the *Beagle*. He devoted an entire chapter to it in *The Expression of the Emotions in Man and Animals*, published in 1872. Blushing is a uniquely human characteristic. But why should such a behaviour emerge? What evolutionary advantage does it bring? The fact that the blush is invisible among dark-skinned people rules it out as an effective sexual signal. The prevailing view in Darwin's time was that the blush was part of God's design to expose human shame – a silly idea that Darwin justly refutes with the observation that it would be unfair then, wouldn't it, to inflict the trait especially on those who are merely shy.

Darwin drew evidence from friends and correspondents about this 'most peculiar and most human of all expressions'. He asked whether children blush, and if they do, but not from birth, at what age they start. He asked whether blind people blush. He confirmed that blushing was not dependent on skin colour by finding subjects in whom scar tissue or albinism allowed the coloration to show through. One eager lady correspondent informed him that women who blush prettily fetch a higher price at the sultan's seraglio. He asked the sculptor Thomas Woolner to report on how low his ingénue models blushed: 'I daresay you must often meet and know well painters. Could you persuade some *trustworthy* men to observe young and inexperienced girls who serve as models, and who at first blush much, how low down the body the blush extends?' The answer was that the *appearance* of a blush is generally confined to the face and neck, although the person blushing may *feel* as if the whole body is blushing. (Thus, the thigh of an aroused nymph might well become *flushed* owing to a similar effect of increased blood flow through her capillaries, but this has a physiological, rather than mental, cause, and so is not a *blush*. Monkeys, too, Darwin noted, 'redden from passion'.)

In the end, Darwin concluded that blushing arises from the human 'habit of thinking what others think of us'. It was not a result he was especially happy about, as it emphasized the uniqueness of human consciousness over our evolutionary connection with other species. But it explained the observations: why infants do not blush, but children do; why the mentally retarded seldom blush, but blind people do; why we tend not to blush when we are on our own, but can nevertheless blush at an embarrassing memory. What it didn't really do was to explain why we find blushing so attractive in others, which, for Darwin, interested as he was in the mechanisms and effects of reproduction, was surely the point. Today, scientists are able to measure capillary blood flow and even the temperature of rosy cheeks, but are still not much closer to an answer.

'Darwinian Man, though well-behaved, / At best is only a monkey shaved!' So sings one of the lady professors in *Princess Ida*, Gilbert and Sullivan's musical satire on feminism, evolution and other novelties thought up to confuse the Victorian paterfamilias. From Darwin's shaven monkey to the naked ape of Desmond Morris we are endlessly reminded about our skin: its vast expanse, some two square metres in all, making it, in answer to a famous trick question, the largest organ of the human body; its colour relative to others, and our curious propensity, having declared that this matters so greatly, to ignore its real hue and settle for calling it 'black' or 'white'; and above all its sheer, vulnerable, embarrassing nakedness.

So great is our sense of exposure that we have developed an elaborate vocabulary to deal with it. The concept of the 'nude', as Kenneth

Clark points out in his masterly (and only marginally lecherous) examination of the subject, was devised in the eighteenth century as a way of enabling artists to work from and talk about the naked body without shame. But with the advent of film and, shortly afterwards, abundant pornography, we need also to distinguish between 'the nude' and 'nudity', and, tediously, between official classifications such as partial nudity, rear nudity, full frontal nudity, brief nudity, natural nudity, sexual nudity, graphic nudity, and so on. We even have the oxymoronic state of 'fully clothed nudity', seen, for example, in a 1956 short film about Lady Godiva in which Maureen O'Hara rides through the streets of a Hollywood Coventry wearing underwear, a full flesh-toned body suit and, just to be on the safe side, hair down to her knees. Tiny differences in usage of these terms have huge semantic implications. To be 'in the nude', for instance, is not quite the same as being naked, yet nor is it the equivalent of an artistic nude. It implies the presence of a spectator whose motives are not primarily aesthetic. And it also carries with it the expectation that the nude is in some way there to be watched. Thus, an actress photographed by paparazzi will conventionally be described in the tabloid press as being 'caught in the nude', whereas a politician photographed in some compromising situation will typically be called merely 'naked'. There are hundreds of academic studies of the nude in art, but comparatively few of the nude in film, in advertising, on the beach or in the bath. Sometimes, we even draw a veil where there is no veil to be drawn. For example, Classical scholars have often translated the Greek *gymnos* and Latin *nudus* to mean 'scantily clad', but the words did in fact mean 'naked', notwithstanding the views of do-gooders such as William Gladstone, who could not believe that naked athletics contests were normal in Homeric Greece.

The essential difference is all to do with context and intent. A naked person may become a nude if painted in oils but perhaps not if photographed, if seen in a studio but not in a club, if holding still but not if moving (a streaker at a sports event is not a nude), if maintaining a certain codified attitude, such as the pudica pose, as we have seen, but not if flaunting their nakedness. The absurdity of these distinctions was pushed to its limits in British strip clubs during the mid twentieth century when it was illegal for a stripper to appear naked if she also moved. Elaborate acts were devised in which the stripper would remove her clothes while strategically hidden behind fans wielded by other (clothed) dancers. At the end of the act, she would stand stock still for a brief moment under the spotlight.

The Victorian critic John Ruskin is famously supposed to have been shocked when he saw his beautiful bride Effie Gray naked on their wedding night. The marriage was not consummated, and the couple later divorced. Ruskin stated that, 'though her face was beautiful, her person was not formed to excite passion. On the contrary, there were certain circumstances in her person which completely checked it.' Effie told her father that Ruskin 'had imagined women were quite different to what he saw I was, and that the reason he did not make me his Wife was because he was disgusted with my person the first evening April 10th'. But why? Some hideous deformity, a birthmark, cellulite? Scholarly speculation, echoed by popular retelling, has it that the great critic was shocked by the sight of the pubic hair so conspicuously absent from the statues that were the habitual subject of his admiration. Matthew Sweet punctures this myth in his book *Inventing the Victorians*, citing Ruskin's previous familiarity with 'naked bawds' while a student, but still does not explain the difficulty, choosing to ignore Effie's guileless description of her normal femininity. Clearly, Ruskin had sensed something about her naked body that was not to his liking. Perhaps it was just the unexpected difference between the warm, breathing, supple flesh of an animate body and the cold marble that he was accustomed to perusing. The feeling, it seems, was not entirely uncommon. For example, Arthur Thomson's *Handbook of Anatomy for Art Students*, published in 1896, makes much of its author's disappointment that female buttocks are not always the smooth globes of Classical statuary. Fat here, he writes, 'is particularly liable to occur in female models past their prime, and imparts a grossness to the form at variance with the delicacy and refinement displayed in earlier life'. Perhaps Ruskin would have been more comfortable with modern porn magazines, in which, in contradistinction to medical publications, there has often been a legal requirement to retouch images to trim away (women's) body hair and in other ways to 'heal' the models on display. Whether it's done for legal or aesthetic reasons, such prurient editing may not produce 'nudes' in the sense understood by the art world, but it does serve to distance the subject from the ordinariness of being merely naked.

It is only human clothing that makes bare skin exceptional and human morality that makes it troubling, as I am reminded when, for the first time in my life, I attend a life-drawing class. I began this book in a dissection room, where I was attempting to reproduce on paper the appearance of dead body parts. As I draw towards a conclusion, it at least feels more natural to be drawing from living subjects.

More natural, but certainly no easier. About twenty of us have assembled at a community centre in the windswept outskirts of Cambridge – two-thirds are women, and there is a good span of ages. We sit on cheap plastic chairs placed in a large circle on a floor marked out for basketball. In the middle of the circle are two young women, who I learn later are university students earning some cash. They have steps to sit on and handrails to grip in order to strike interesting poses. Without fuss, they shed their dressing gowns and get into positions as directed by the class instructor. We each choose one of the models and begin to draw. Immediately, I am in all sorts of trouble. I find it difficult to get the major proportions right between the torso and the limbs. My pencil creates hard, sharp lines that fail to communicate the softness of the skin and the diffuse fall of shadows across the body. It gets worse when I try shading, and my lack of technique is ruthlessly exposed. As the evening wears on, though, I feel I am discovering one or two tricks, such as extending a line beyond what I see in order to give a sense of movement and life to the muscles. The mere creation of a drawing, however poor, seems to produce a connection with all art. There are aspects of my paltry sketches that recall ancient heads and figures. The two women, standing naked in front of us, on paper have become, through no fault of theirs and no great skill of mine, nudes.

The second time I go along, one of the subjects is a stocky, muscular man, who is introduced to us as Andy. He has been asked to lie on his back with his head dropping low. He looks extremely uncomfortable, although he appears to be on the verge of dropping off to sleep. Oddly, he has a white bandage across the bridge of his nose. It is not clear whether this stems from some injury or has been placed there for artistic interest. Our instructor, Derek Batty, invites us to draw his face in this upside down position – 'an interesting psychological challenge'. He's referring to what is known as the Thatcher illusion. In 1980, Peter Thompson, a psychologist at the University of York, demonstrated the prime importance of the eyes and the mouth in facial recognition by taking a photograph of Margaret Thatcher, Britain's then new prime minister, and altering it by inverting only these features. When the altered head is seen upside down, it is easy to recognize who it is because the eyes and mouth appear correctly. But when the head is the right way up, with the eyes and mouth inverted, it appears monstrous. I note with amusement from his paper that Thompson thanked the York Conservative Association for supplying the 'stimulus material'. Any way up, I find, a face is much harder to draw than a body.

After the classes, I stop the models and ask them how they feel that we have been granted this exceptional permission to stare intently at their bodies and faces. The thing that surprises them, they tell me, is how unaware of the class they quickly become. The nudity is not an issue. Their minds are elsewhere. Andy is psyching himself for tomorrow's championship kickboxing match – which at last explains the bandage. The woman I have tried to draw, Rosie, passes the time thinking about her PhD thesis (on Soviet cinema). But, she adds, 'if Derek mentions a body part, I immediately feel a need to move it.' Her remark reminds me of Darwin's exploration of blushing, which he regarded in the end as an involuntary response to another's focus of attention on the body.

Our skin, all two square metres of it, about the area of a single bed sheet, is a screen. It carries the projection of who and what we are, like a film in a cinema. It is also a screen in another sense – like one that stands in the corner of a room, blocking the view and offering protection to the body on the other side. Biologically speaking, the skin is a formidable membrane between solid and air, between our innards and the world outside. In its depth lie the sensors by which we feel pleasure and pain, and our means of defence against many infections. And yet in cultural terms, the skin is the thinnest of barriers between interiority and exteriority. Its thickness counts for nothing when our health, our age and our race are displayed for all to see on its very surface. The skin is both our self-protection and our self-revelation.

This duality is at the heart of its meaning. Before modern medicine, the skin was seen as the guarantor of corporeal integrity, not so much part of the body as its appointed gatekeeper. To an extent, it was even regarded as dispensable; perhaps it was a barrier to the enlightenment of the self within. In the Bible, Job escapes the suffering to which he has been subjected as a test of his faith 'by the skin of my teeth' and rejoices: 'after my skin is destroyed, this I know, / That in my flesh shall I see God'. Yet, to other ancient writers, the skin also comprised at least part of that self. In *Metamorphoses*, Ovid tells how the satyr Marsyas, flayed alive after being defeated in a contest with Apollo, begs: 'Don't rip me away from myself.' Here the skin is the organ of our literal self-possession. It holds the rest of us in place. The ambiguous status of the skin – is it *of the body*, or is it a kind of disposable wrapper *for it*? – perhaps reflects broader unease at the whole notion of human embodiment that is bound to spring from the dualistic idea of body and soul.

These perceptions of the skin had important medical implications. Many diseases were not understood, as we would understand them now, as diseases 'of the skin', but were regarded as signs on the surface of bodily (and moral) decay below. Leprosy is especially abhorrent in biblical accounts. The book of Leviticus contains a long and almost clinical description of the various ways in which the disease might appear on the skin, and the precautionary measures to be taken according to the extent of skin affected and, crucially, whether the infection appears more than skin-deep, ranging from quarantining the patient to obliging him to shout out: 'Unclean! Unclean!'

But while the skin may advertise diseases such as leprosy, smallpox or syphilis, it obscures the presence of others. The skin is opaque to us all. Unable to see past it, even expert physicians are apt to make startling misdiagnoses. Appendicitis was typically not identified until the patient began vomiting fecal matter, for example. One recommended treatment for the abdominal pains that were among the early symptoms was to eat quince, which was only likely to exacerbate the condition. But similar difficulties confront doctors today. A friend of mine, complaining of intermittent hearing loss, was seen first by a neurologist who suspected vasculitis, a disease that destroys blood vessels; she was also tested for syphilis and then put on a course of steroids, which proved ineffective. A second neurologist favoured multiple sclerosis, but tests on her epidural fluid proved negative. A series of hearing experts then stepped in, the third of whom discovered at last that all three bones of one middle ear had fractured. These were then surgically removed and replaced by metal prostheses. To be fair to the medical profession, the rest of us, too, use the skin as a convenient curtain to deny the messiness of what goes on beneath. Norbert Elias's *homo clausus*, man 'severed from all other people and things "outside" by the "wall" of the body', has become a touchstone of the human condition. In cartoons, for example, convention demands that physical blows bounce resiliently off the body – or temporarily flatten it; they do not actually rupture the skin. We think to seal ourselves off against the world.

The psychological impenetrability of the skin – even to surgeons who hesitate to bring down the scalpel for fear of making matters worse (Hippocrates: first do no harm) – remains one of the most unshakeable truths about the human body. It explains the high value that we place on anything that gives, or appears to give, a picture of what is going on beneath – the humours, the phrenological head, the X-ray, the genetic profile, the ubiquitous 'scan', to which we gaily refer without regard for its technological means, or for that matter its diagnostic power, almost as if it were a modern miracle.

If the skin is a screen, then what's on? The film of life begins as a blank. Innocent skin is 'smooth as a baby's bottom': unmarked by disease, sin and the ravages of time. But how long can it remain so? The smoothness of the bottoms and the rest of, let's say, Antonio Canova's sculpture *The Three Graces*, an early nineteenth-century marble statue famed for its coolly erotic beauty, was not only a statement of artistic prowess but also a reaction to the ugly reality of the 'rotting, eruptive and squamous skins that constituted the actual bodyscape in the eighteenth century'. The smoother the skin, the more impenetrable and therefore protective of the underlying body it seems. Unction – the consecrating anointment with oil of a priest or a monarch – smoothes the appearance of the skin, producing a glossy sheen that more sharply defines and, in a sense, hardens this barrier, sealing off these leaders from their unclean subjects. The application of suntan oil contains a secular echo of this ritual, sealing the body against damaging solar radiation. The oiled muscles of the bodybuilder, the rubber and leather sheaths of the fetishist, and the shiny, chrome bodies of CGI adventure heroes all aim, for their own particular reasons, to produce the same hermetic seal.

Conspicuous areas of bare skin may be a sign of vulnerability – Adam and Eve in the Garden of Eden, Christ on the Cross, Hans Christian Andersen's fairy tale of 'The Emperor's New Clothes'. But they may also be an assertion of power: Lady Godiva wins her fellow citizens a tax break in return for her naked gallop. The bare chest of the Russian prime minister Vladimir Putin has become such a political phenomenon that even the *Journal of Communist Studies* has been driven to comment on it. Frankly, I'm baffled as to how to react to this. Am I supposed to admire him, fear him, fancy him? What if Prime Minister 'Dave' Cameron stripped off his shirt? How would I feel about that? Knowing him as we do, we perhaps interpret Putin's naked torso as an expression of his authoritarianism, and yet the figure of Liberty, too, in Eugène Delacroix's painting *Liberty Leading the People*, is defiantly barefoot and bare-breasted. (Similar displays can lead to a curtailment of liberty today, however: in 2003, an Australian Member of Parliament was ushered out of the debating chamber for breast-feeding her baby, supposedly having infringed its rule of 'no strangers in the house'. As the Australian cultural historian Ruth Barcan commented: 'it was not so much the baby as the breast that was the stranger in the house.')

The skin is a medical message board as well. *An Essay Concerning the Infinite Wisdom of God, Manifested in the Contrivance and Structure of the Skin* by 'A Lover of Physick and Surgery' is a typical early modern mixture: well-observed description of the body combined with frequent reminders of its divine ideality. Each chapter concludes with an incredulous attack on atheism. All body parts are just the right size and shape, notes the author, and there is much moralizing speculation that it would all have gone horribly wrong for humanity if anything had turned out otherwise. It is because our skin is naked, for example, that our nails are so useful for scratching. That the nails are transparent, the anonymous eighteenth-century writer continues, makes them the perfect thing to indicate the true colour of the blood beneath. They are like little windows through the skin, or indicator lamps on the ends of our fingers, that turn pale with ague, red with 'plethora' or high blood pressure, yellow, green or black with jaundice and other complaints.

The skin may also carry our own advertising messages in a very literal sense. As the skin is still used with such blithe confidence by those in authority to assign us to particular racial groups – London's Metropolitan Police, for example, aspires to describe people of mixed ethnic origins under such muddled terms as 'Asian & White' – so it follows, to some anyway, that new marks might be applied to skin in order to create new categories of social distinction. Historically, marking the skin has often had a quasi-legal aspect to it – from the brands burned into the skin of slaves by slave-owners to the scars left by the lash, marking a person permanently as a criminal. This custom persists today in the benign form of rubber-stamping the back of the hand for entry to a night club, for example. But it is the idea that one might purposely choose to mark oneself that is so notably in the ascendant right now. Skin has never been more on display in Western society than it is today, and never more subjected to our own alterations – adaptations that are designed to communicate a new version of ourselves.

My publishers have recommended that I visit a tattoo artist whom they have previously commissioned to produce book covers. These are not covers of human skin, I should add, although that practice has not been uncommon in the past, especially for binding criminal records and medical works. One Russian poet even bound a volume of sonnets to his mistress in skin salvaged from his own leg, which he was having to have amputated.

The studio – 'parlour' seems an outdated term – is called 'Into You'. It seems well named, with its piled-up suggestions of penetration of the body, physically, through the skin with needles, but also sexually and emotionally. Here I meet the proprietor, Duncan X, whose name, changed by deed poll, is itself a mark. His body is covered with bruise-blue designs: skulls, coffins, various slogans, a masonic symbol of some sort high on his forehead. His face is mostly clear except for a couple of teardrops coming from his left eye. He also has his mobile phone number tattooed on the back of his hand, a useful reminder for him, and a reminder to me that we all occasionally write on our skin for such purposes.

For Duncan, the individual motifs are not as important as the overall pattern, here lighter, here denser, broadly symmetrical, but with smaller asymmetries and chaotic details, just like the human body itself. 'It was important it wasn't a picture,' he says of his first tattoo, gained at the age of twenty-one in an attempt to shock his parents, who were both medical doctors. 'It was the tattoo concept, the ultimate rebellion.' From there, he continued to cover himself. 'I would feel really strange if I didn't have them. They are like armour, I feel protected, but it is also like having the skin scraped off to reveal the real you.'

Duncan's remarkable work includes inspirations from woodcut medieval maps and paintings by the Bruegels as well as more conventional motifs from the culture of tattoo traditionally prevalent among seamen and prison inmates. The people who seek him out are not the kind who want a tattoo – or the transferred-on illusion of one – for reasons of fashion. 'My customers are more concerned about changing themselves. They are people in a state of change, and this is a very visible one. I have seen people liberated,' he tells me. He does not see it as part of his job to ask why a person wants a particular design or what some foreign script means. The psychotherapy is in acquiring the mark. As in supposedly remote or primitive cultures, a tattoo marks a rite of passage. There are plenty of reasons *not* to be tattooed: the permanence, the time-consuming process of getting it done, the pain involved, the breaking of the skin. All these barriers become part of the rationale. 'They will have thought that these are not strong enough reasons to stop them going ahead.'

For these people, as for those who engage in the cutting medically classified as 'self-harm', and even perhaps for some who undergo cosmetic surgery, pain is an essential part of the experience. These actions seem to be secular versions of the mortification of the flesh. Mortification of the flesh, which is a traditional feature of many religions, may take a number of forms, most commonly degrees of fasting, but more extreme forms involve the creation of visible scars through actions such as self-flagellation or pulling on strings tied to hooks in the skin. Pain is experienced as an emphatic aspect of the self-denial of normal pleasures, while the scars are the conspicuous public sign of the celebrant's piousness. In today's secular equivalent, these things seem to reflect a desire for felt existence in a world where the regulated environment of civilization does so much to numb our senses, and to be a desperate assertion of identity, transforming the skin that we have been given by nature and that is recognized by authority and reinscribing it as our own. The skin is, as always, our most sensitive means of interaction with the world, and yet somehow still seems to be the barrier to our deeper immersion in it.

With this, we sense that we have reached a kind of limit. We stand at last on the shore of our island selves. And yet. 'Why should our bodies end at the skin?' the science historian Donna Haraway demands to know in 'A Cyborg Manifesto', a polemical appeal to reimagine our existence free of the fetters of gender, race and all the other social conventions that advertise themselves on the fleshy surface of the human body. Haraway notes that the shell has already been breached: we already invite into our bodies 'other beings encapsulated by skin' through xenotransplantation of tissue from animals such as pigs and monkeys and even injections, such as of the botulinum bacteria used in cosmetic Botox treatments. These dermal forays can be read as signs of our urge to explore beyond the boundary of the skin. Is *homo clausus* finally opening up? If so, what delights – and what dangers – await us? These are the possibilities we shall explore in the final chapter of *Anatomies*.

PART THREE

The Future

Extending the Territory

What is it that the kids from *Fame* sing? 'I'm gonna live for ever / I'm gonna learn how to fly.' It's not that they really plan to do either of these things, of course. For them, it's more about the feeling of being in the physical, performative moment. And yet, there are, somewhere deep in us, those earnest wishes. We admire what the body can do, and still wish it could do more. We dream of extension – of our physical capabilities, of our sense perceptions, of the span of our brief lives. Curiously, this desire is directed chiefly towards our corporeal selves. Our minds are untouched by it; for some reason, we do not yearn for greater wisdom or imagination in quite the same way.

Such dreams are not new. We may be created in God's image, but our gods we imagine as super-able versions of ourselves. Lakshmi, the Hindu goddess of prosperity, has two pairs of arms, while Brahma also has four heads. Besting both of them, Guanyin, the bodhisattva of compassion of East Asian Buddhism, has eleven heads and a thousand arms. The Greek fertility god Priapus and his Egyptian counterpart, Min, have permanent erections. The Greek mother goddess Artemis sports multiple breasts.

The *Metamorphoses* of Ovid heads a vast literature demonstrating that the human urge to improve or transform the body, or to exchange one body for another, is both strong and constant. The theme continues through powerful stories such as Mary Shelley's *Frankenstein* and the fairy tales collected and added to in the nineteenth century, such as the Brothers Grimm's 'The Frog Prince'. Today's Hollywood blockbusters have revived the genre with the help of realistic computer graphic imagery. The character transformations in these stories may be offered as a salutary lesson or moral to the audience, as with the appearance of the stone guest in the Don Juan legend, whose unexpected movement warns that the Don will not go unpunished for his sins, or they may be personally liberating and capable of alter-ing social perceptions, as in the *Shrek* films. Either way, they are life-changing events.

All technologies are, in Marshall McLuhan's famous dictum, 'extensions of man'. Often, it seems, our desire is for greater powers of destruction. When we dream of extending the capabilities of the hand, for example, it is often a weapon we wish to add, as we are reminded when we see a child blow the imaginary smoke away from the finger he has just used to shoot his friend. 'My right arm is com-plete again,' exults the murderous barber Sweeney Todd as he wields his beloved razors in Stephen Sondheim's musical. But a similar tech-nological extension to human capability serves a more benign purpose in *Edward Scissorhands*. Tim Burton's film derives from tradi-tional archetypes such as 'The Sorcerer's Apprentice', in which inventors create mutant living creatures, and in particular from the German story of Struwwelpeter, a cautionary tale about a boy who never cuts his nails or combs his hair. The story follows the conven-tional path as Edward is initially misunderstood before performing wonders and finally being accepted for what he is. It shows how physical extension can lead swiftly to a more complete personal transformation.

Whereas Ovid relied on changes in natural types, metamorphosis now takes a technological form. Both natural and artificially assisted transformations, however, show our commitment to the never-end-ing invention of our own bodies. With the rise of biotechnology, we can expect to see a convergence of these two worlds, the mechanical and the organic, and a closer integration between our natural bodies and the features with which we extend them.

As McLuhan observes, our technological extensions demand our obeisance. Our bodies must become their servant if in turn they are to be useful to us. I am curious to learn how this works in real life. To find out, I have arranged to see Jody Cundy, a multiple gold medallist in the British Paralympic team. He was formerly a world-champion swimmer but now enjoys a no less exalted status as a cyclist. At birth, his right leg had no ankle or foot, finishing with two toes at the end of his shin bone. He now uses a variety of prosthetic legs, with special high-performance versions made of carbon-fibre for use in competition. This is one degree of extension. Jody's bike, also made of carbon-fibre, is another. It is his body plus his artificial leg plus the bike that together are able to achieve record-breaking speeds. I am intrigued to know where 'Jody' stops and technology takes over.

I arrive at the National Cycling Centre in Manchester, where the athletes are in training for the 2012 Paralympic Games. A large banner outside the velodrome reads: 'Chasing immortality'. Jody has tousled strawberry-blond hair and an uncomplicated, outgoing manner. It is no surprise to learn that when he is not on the track he earns a living by giving motivational talks.

Jody was fitted for his first prosthetic limb at the age of three. Then, every six months as he grew up, a new fitting would be required. At first, these were elaborate metal contraptions which had to be strapped to the thigh by means of a kind of leather corset and around the waist with a belt. 'Dad would have a toolkit of stuff to fit the legs,' Jody remembers. Today's attachment is a great improvement. It has a socket custom-shaped to match the tapering stump below Jody's knee and a lubricated silicone liner to create an airtight seal. 'The only time I've had a leg I don't really feel is with these latest ones where the fit is so good,' he says.

Jody started cycling as part of a regime of complementary fitness training for his swimming. But then one day a coach spotted him going round the track and thought he looked a natural. He made the difficult decision to change events and never looked back. 'I went from complete novice to standing on a podium in about eighteen months,' he tells me, as he nonchalantly changes his 'walking' leg for a cycling version, which has an integral clip to lock it into the cycle pedal.

After a brief chat with his coach about the programme for the day – perhaps to try a few starts and a few accelerations – he is set off on forty laps of warm-up. A motorcycle sets the pace, and the cyclists follow close behind in its slipstream. Jody turns in lap times of twenty-six seconds. The pace seems leisurely, but I calculate that he is already moving at more than thirty kilometres per hour. By the time the last lap comes round, he'll be doing sixty. During what the athletes call an 'effort', he can reach seventy kilometres per hour.

Jody's normal left leg is exceptionally well developed, as one would expect for a competition cyclist. His calf resembles a plump ham. The prosthetic limb next to it may be sculpted to look like a natural leg (though without the exaggerated musculature), but it does not work quite like one. The differences mean that Jody must use his body differently from other cyclists, and must think differently in order to produce the required actions. A track cyclist normally uses the hinge of the ankle and muscles in the lower leg to bring the pedal back up from the bottom of each circular orbit (the foot being strapped to the pedal). Because his right leg lacks a normally pivoting ankle, however, Jody must instead use a group of muscles in his hip (collectively known as the iliopsoas) to produce this lift. The prosthesis does not give him any power advantage. If anything, Jody feels that it is his normal left leg that is inexhaustible, simply because it is always the right that gives out first, limited by the strength not of the muscles in the calf but of the quadriceps in his upper right leg. Laboratory tests show that, although it tires first, Jody's right iliopsoas is in fact stronger than his left because of the particular action required of it to compensate for the fact that he has no muscles at all below the knee.

For most of us, cycling is something that we do without thinking. But Jody has to think about it, both in order to improve his performance and in relation to his disability. When he pushes down with his left leg, Jody explains, 'you've got this whole unit that wants to do stuff. Whereas on the right, I've got this motion' – he awkwardly hinges his hip and lifts up his thigh. 'It's almost as if I'm trying to hold on to the inside of the leg when I come up. The hardest part is getting through the bottom of the pedal. I struggle with top and bottom dead centre. I don't feel I am getting any power.' These are the points in the cycling action where the ankle would normally hinge and the muscles of the lower leg would be most busy. This is especially important at the start of an event. In training, Jody's strategy for this is to 'trick the body into learning to do it quickly', and then to be able to repeat the trick as he moves progressively on to harder exercises, which he does by repeatedly swapping the single gear on his track bike for gears of a higher ratio.

Jody's main event is the one-kilometre time trial, for which he won gold at the Beijing Paralympic Games, in a time of 1 minute, 5.47 seconds. It is a difficult distance for physiological reasons, being long enough that the body begins to suffer, and lactate, a product of the breakdown of glucose that provides energy, builds up painfully in the muscles. Because so much blood is drawn to his legs during the event, Jody finds he has to lie down immediately afterwards in order to restore his balance. 'You can't imagine doing it much more without passing out,' he says with feeling. His remark reminds me of Emma Redding's dance-based exploration of the point of exhaustion.

Jody's sense of his own body alters when he is cycling. Normally, his body envelope is defined by his natural biology: it ends where his body stops, at the literal limits of his physique. His left leg stops at the toes, but his right stops just below the knee. But when he's wearing his artificial limb, which weighs a lot less than a lower leg of flesh and bone, he says he feels that weight disproportionately because it is an inert attachment; it makes his leg as a whole feel a bit like a pendulum. At low speeds, Jody can feel the difference in his legs. When he is cycling at speed, however, his body envelope expands to include the carbon-fibre prosthesis and even his bike, which weighs just under seven kilogrammes. 'It never feels like something's on the end of the stump,' he tells me. 'And with the leg being made of the same material as my bike, you have this feeling of the leg and the bike being one and the same. You feel it most when you're accelerating, and the sensation of all the force that's being applied to my prosthetic leg is making its way seamlessly to the back wheel. It's an amazing feeling.'

It had been convenient for me to use my own bicycle to get from central Manchester to the velodrome. As I ride away, the sun is shining. My less-than-fit self and my low-tech bike hardly add up to a harmony of man and machine. My speed is a modest fraction of Jody's. The experience for me is more about being out in the air, moving with freedom through the cityscape, an extended capability as close to flying as most of us ever get without artificial power.

In the promised era of self-transformation – biological, technological, psychological, chemical – how do we really feel about the extension of our own body's capabilities? Should extension be clearly artificial, or should it ideally be indistinguishable from the host body, united in one integral organism? Before we take a position one way or the other, it is perhaps worth remembering that the distinction is already far from clear. As one bioethicist drily notes, even those prone to object that we would no longer be our natural selves through such intervention tend to be 'folk who wear eyeglasses, use insulin, have artificial hips'.

One of the most prevalent images we have of entities that can both fly and live for ever are angels. Where I live in East Anglia, they are pinned in their dozens to the roofs of its great churches like butterflies in collectors' drawers. They represent a state of being that is inaccessible to embodied humans, yet so obviously one that we would like to inhabit and experience for ourselves. The means of attachment of the wings seems to reflect this ambiguity. In strictly anatomical terms, it is hardly viable. The wings usually sprout from the shoulder blades – perhaps the jutting part of this bone suggested to the artists who first developed such images a missing avian appendage – but there is never a hint of the bulky musculature that would be necessary to drive them. They represent the *idea* of flight, but no realistic prospect of it.

Wisely, artists seldom choose to depict angels actually flying. Indeed, the Bible provides only one such glimpse (when Daniel sees 'the man Gabriel . . . being caused to fly swiftly'), and in general is ambivalent about angels' need for wings at all. When paintings and sculptures show wings, they are clearly borrowed from birds and magnified proportionally. As extensions of man, though, they fail every practical test because the artists never augment the bone and muscle in a way that makes physiological sense. Instead, they should really be seen as emblems of divine power. As the author and Christian apologist C. S. Lewis observed: 'Devils are depicted with bats' wings and good angels with birds' wings, not because anyone holds that moral deterioration would be likely to turn feathers into membrane, but because most men like birds better than bats. They are given wings at all in order to suggest the swiftness of unimpeded intellectual energy. They are given human form because man is the only rational creature we know.'

Whereas angels have human form and superhuman powers, robots are technological devices with human powers. After all, they are designed for the most part to do jobs that we would rather not. But to carry out human tasks does not necessarily require a human form. Strange to find, then, that in the burgeoning robotics research community, there is still an unaccountable fondness not only for modelling these devices very literally on what humans can do, and the way they do it, but also for giving them human likeness. I read, for example, of projects to create robots that will be able to push wheelchairs. This seems to miss the point: surely the answer is an 'intelligent' wheelchair, rather than a conventional wheelchair with a second human-like machine to push it along. In Karel Čapek's *R. U. R.*, I might add, the robots take on human form, but only because their creator 'hadn't a shred of humor about him'. In our literal way, we are driven to fashion both angels and robots in our likeness because human form provides the most compelling vehicle for describing human aspirations.

At the moment, robots amuse us because they look so unnatural in their mimicking of our actions. In the future, if the technological dreamers get their way, they will look so human that it will be no laughing matter. Uncanny Valley is the place where humans begin to feel genuinely uneasy at something's ability to appear human when it is not. This 'valley' is, in fact, a trough in a line graph that plots human enthusiasm for robots against their increasing human likeness. The line begins high while it is perfectly clear that robots are just machines. But shortly before they become so realistic that we can no longer tell whether they are human or not, there is a dip – a stage when they simply appear very creepy. Other strange creations already populate Uncanny Valley – Ron Mueck's *Dead Dad*, for example, with its pallid 'skin' and body hair, or the ultra-realistic dolls known as 'reborns' that women sometimes carry as a substitute for a baby that has grown up or one that never arrived. We are fast approaching the point where we will have to decide whether we are going to pass through Uncanny Valley and increasingly share our lives with such creations or turn back.

The Geminoid series of robots created by Hiroshi Ishiguro at Osaka University is perhaps at the pinnacle of human resemblance in robotics. Ishiguro's latest version is fashioned after a Danish colleague, Henrik Scharfe, and comes complete with skin, hair, blinking eyes and a stubbly salt-and-pepper beard to mask its metal innards. Scharfe's own recent published research examines how trust may be built in encounters with his mechanical alter ego. Such innovations may represent a departure from our comic-book expectations of what a robot should look like, but it is important to remember that robots were not originally envisaged as shiny metal helpers with square limbs, red eyes and wheels for feet. Nor did Frankenstein's monster ever have a bolt through his neck. The first illustrated *Frankenstein*, an edition published in 1831, thirteen years after the original, shows the creature stunned and stupid-looking, but with perfect human musculature. The suggestion is of biological life, not some crude mechanical reassembly.

In general, technology has a habit of realizing our dreams in ways other than we imagined it would. We wish to fly? We don't grow angels' wings. We log on to Google Earth instead. Even an artificial heart looks more like a piston engine than a real heart. I was startled during one of my anatomical drawing classes when I caught sight of a piece of plastic pipe lodged in the tangle of blood vessels within the heart cavity of one of the bodies. The straight line and even colour of this surgical insert contrasted so starkly with the variegated colours and textures of the surrounding tissue.

Angels and robots help us to think about where the boundaries between human and nonhuman (or extra-human) truly lie. So, which is it to be? Unashamed technological add-ons, such as Jody's limb? Technology made to look biological, such as robots with stubble? Or biology all the way? Our choices will depend on what we feel comfortable with, or perhaps on what we feel least uncomfortable with. It is notable that potential organ recipients currently prefer the idea of mechanical devices, while their surgeons prefer xenotransplantation – the transplanting of organs or tissue from nonhuman species. They like it because it allows them to continue working in the familiar medium of biological tissue. If the rest of us are disinclined to go along with them, it may be that the medical profession has only itself to blame, as we shall see in a moment.

The hybrid species that began to appear in decorated manuscripts, in bestiaries and as gargoyles during the medieval period – which included human features, especially arms and hands, as well as human eyes and faces, along with desirable animal attributes such as wings and tails – were not simply fanciful renditions of exotic species based on Chinese whispers, nor the excited celebration of biodiversity that we might easily take them for today. Instead, these hybrids between man and the animals sought to comprehend changes in man. Fantastical physical transformation was the pre-modern world's way of exploring and coming to terms with actual psychological transformation. It is key to our understanding of these images to know that while a character's external appearances may change, his or her identity is preserved. It is the same person, only in a different guise. The new guise reveals the new psychological state. It is the same as in Ovid's *Metamorphoses*. When Jupiter rapes Io, and Juno then punishes Io for adultery by changing her into a white heifer, she is still beautiful in her way, but is now revealed in her bestial character. She is still Io, and can recognize her father, but is sadly unable to tell him it is she, except by means of her cloven footprint, which leaves the letters IO in the earth. In Homer's *Odyssey*, Odysseus's men waste a year on their voyage back to Ithaca feasting at the house of Circe, who transforms them into pigs. In appearance and behaviour they are pigs, but their senses and memories are the ones they had as men.

Strict rules apply in tales of metamorphosis. Without them, it would not be clear quite what degree of change we should regard as remarkable and worth a story. These rules also provide a framework for moral philosophy. If a werewolf is, as we have seen, a man (with human eyes) in a wolf's body, then he has the rights and duties of a man. Is killing a werewolf then to be considered homicide? Is a human-eating werewolf a cannibal? Reimagining a dramatic human encounter in ways like this may help to resolve a dilemma about how to deal justly (for the period) with the psychologically disturbed person who has committed a terrible crime, for example.

If psychological disturbance is one predicament where body and mind may no longer be in alignment, then xenotransplantation is another. In 1984, four-year-old Baby Fae received the heart of a baboon in an operation at the Loma Linda University Medical Center in California. The procedure was quickly decried as 'improper' and 'unnatural', although it was to pave the way for successful human organ transplants in children. The Ovidian rule here was that a baboon is sufficiently similar to a human child that the transplant operation was biologically worthwhile, and yet not so similar that sacrificing it for the purpose counted as murder.

It does not help to put us at ease, however, that the chosen animal for many surgical procedures is increasingly the pig, a creature that, as Homer reminds us, comes with an all too familiar cultural back-story. The animal reminds us of ourselves at our worst, with its gluttony and its promiscuity, and its naked, fleshy appearance. Scientists favour pigs over other species because they are close in size as well as in some important immunological respects to humans, because they breed rapidly, and because, being reared chiefly for food, they are less strictly regulated than other candidates such as apes and monkeys and raise fewer ethical qualms. In short, the pig taboo is weaker than the ape taboo. This preference seems 'altogether peculiar from a lay point of view', according to the medical anthropologist Lesley Sharp, because we also associate pigs with filth and defilement. If pork is still subject to dietary prohibitions in many religions, how can we think of inserting pig tissue permanently into the body? The pig's very suitability in bio-logical terms – its relative closeness in some respects to humanness – is also its problem in cultural terms.

In order to persuade the relatives of a human donor to consent to a transplant, the promotional message is often that 'the lost loved one can "live on" in others'. It should come as no surprise, then, that people start to wonder just how much of an animal 'donor' might also 'live on' inside them. Research surveys yield some lively responses. One subject observed that it would be 'a little strange' to have the heart of a baboon. 'Would I start baring my teeth and bottom?' It is no surprise, either, to find that patients happy to discuss their heart-valve surgery often omit to mention the pig that is the source of the replacement valve.

If our fabled enthusiasm for crossover with other species seems to have deserted us just as it becomes a medical possibility, it may be because science has done itself few favours. We have seen ample evidence of medical pioneers' readiness to use all manner of human and animal subjects in transplant experiments through the ages. But perhaps the most notorious modern innovator in the field was the pioneer of the 'monkey gland' treatment, Serge Voronoff, a figure whose bizarre exploits inspired some excellent satirical fiction, a song by Irving Berlin and a lethal-sounding cocktail of absinthe and gin.

Voronoff was born in 1866 in Russia, and pursued his investigations during a long career as a surgeon in France. But his inspiration came from Egypt. During an extended sojourn there in his thirties, he 'made a great number of personal observations on castrated men'. These eunuchs looked prematurely aged in his judgement and seemed on the whole to die quite young. It was, he thought, 'something more than a mere coincidence' that men not so impaired continue to be sexually active in old age.

Voronoff reasoned that if he could graft tissue from young men's sexual organs into old men, then it might prolong their life. There was 'no question' of his obtaining human testicles – that would be 'a mutilation', he noted with perhaps just a tinge of regret – but since livestock are often castrated, there was always 'material'. He made his first experiments on goats and bulls, cutting their testicles into half-centimetre slices, and then introducing these into recipient animals' scrotums. Slices were used in order to increase the surface of contact between donor and recipient tissue, thereby promoting vascularization, the formation of blood vessels necessary for the graft to take. The animals generally survived. Voronoff's 1926 memoir proudly shows photographs of a bull named Jacky and the offspring he was responsible for producing after the transplant.

Before any actual prolongation of this animal's life could be observed, however, Voronoff had moved on to human subjects. In his memoir, he rues the fact that volunteers are not allowed by law to donate single testicles – the remaining testicle would in fact come to do most of the job of both, just as a kidney donor's remaining kidney does, and even as one half of the brain can do if the other is damaged. Instead, while the occasional unlucky accident may yield a windfall, he is dismayed to find he must have 'recourse to apes'. In December 1913, Voronoff had successfully grafted a thyroid gland taken from a chimpanzee into a child with a hypothyroid condition. Six months later, he was triumphantly able to bring the child before the French Academy of Medicine. 'Thanks to his graft all the symptoms . . . had disappeared and the child, which was previously so backward as to be almost reduced to the animal level, had recovered his intelligence and his normal growth,' Voronoff wrote later. 'The proof of this statement lies in the fact that four years later, when eighteen years of age, young Jean, whom I had known in 1913 as a poor little imbecile, having but a rudimentary brain and the body of a child of eight, was found suitable for military service and accomplished his duty in the trenches most gallantly.' Emboldened by this success, Voronoff over the next decade carried out hundreds of grafts into human subjects of sex glands from apes, as well as at least one using human testicles. He also tried ovarian grafts in women, inserting the monkey ovaries by preference into the outer labia of the vagina, in an effort to restore hormonal function, if not the full capacity for ovulation.

By his own account, the method was a triumph. In 1923, for

instance, an eighty-three-year-old English gentleman benefited from Voronoff's surgery, 'in spite of the fact that he had the reck-lessness to leave my nursing-home at Auteuil half-an-hour after the operation, in order to get back home by motor-car.' By the time that Voronoff was recounting these achievements, the man was eighty-five and, to judge by the before-and-after photographs, in better condition than ever. Another English patient appears looking slumped and fed up in a picture taken when he was seventy-four; at seventy-seven, he is seen running in spats towards the camera.

Voronoff's moment passed, however, and his death some thirty years after these experiments went almost without notice. He lives on in fictional creations such as the ambitious Dr Obispo in Aldous Huxley's novel *After Many a Summer*, who hopes to exploit the longevity of the carp to prolong the life of his Hearst-like Califor-nian employer, and the Moscow professor Preobrazhensky in Mikhail Bulgakov's *Heart of a Dog*, who implants human testicles and a pituitary gland into a stray dog. The dog swiftly acquires the worst characteristics of both dog and man, thus satirizing the behaviour expected by the Communist Party of the 'New Soviet' citizen.

Serge Voronoff's desperate mission reminds us that perhaps the great-est human extension of all would be an extended lifespan. Who would not opt for a few more years – or decades – of healthy life?

There are two powerful forces behind this thought, one of attrac-tion, the other of repulsion. The first is the alluring prospect of continuing the increase in longevity enjoyed by humankind since the advent of modern science. The age at which we can expect to die has tripled during the course of human history. By 1750, a Swede (the Swedes have kept the best historical records of mortality) could expect to live to the age of thirty-eight. Since 1950, Americans have added an average of nine years to their lives. In Britain, life expect-ancy increased by almost two whole years during just eight years of the last decade. In most of the developed world, life expectancy now hovers around eighty years. This rate of increase is relatively constant, and there is debate about when – or whether – it will top out.

The second factor is of course the spectre of death. As the American surgeon and writer Sherwin Nuland observes, nobody these days is allowed to die simply of old age. National government health departments and the World Health Organization keep statistics that require a cause of death to be given in all cases. 'Everybody is required to die of a named entity.' It is obvious that this data is useful to healthcare planners and actuaries, who need to know the risks of mortality from medical conditions and accidents. But *all* deaths? What really underlies this drive to ascribe cause? What does knowing a cause of death *compensate* for? What does it say about the way we deal with death? Surely its effect is to make us think of death as an accident, as something that might be forestalled – perhaps even avoided altogether – if only we are careful enough. To die at the age of eighty-five, let us say, might seem to demand little explanation. And yet to die at the age of eighty-five as the result of complications after a fall – which happens to be exactly how Serge Voronoff met his end – invites a host of questions. How did he fall? Might that fall have been prevented? What were the complications? Were these avoidable? What if he hadn't fallen? How much longer would he have lived?

Today's visionaries aren't satisfied with extending life by a little bit. They want to extend it by a lot. And they believe they are on the verge of having the scientific tools to do it. Their approach is no longer one of eking out a year here and a year there based on what they can extract, intellectually or surgically, from exceptionally long-lived humans or animals. Their thinking is altogether bolder, and arguably hostile to the traditional philosophy of biology. In short, they regard death as a technical failure. Their project is to identify the causes of this failure, and then to devise the means to eliminate those causes one by one. For this, they have earned the label of transhumanists or, more specifically, immortalists.

The most colourful and controversial of this new breed of thinker is Aubrey de Grey, who is the co-founder of the SENS Foundation. SENS stands for Strategies for Engineered Negligible

Senescence. De Grey formerly worked at Cambridge University's department of genetics, something that has lent his project an appearance of credibility that isn't entirely its due. He is in fact a computer scientist, and he worked at the department in that capacity, only becoming interested in genetics when he married a geneticist there.

We meet in a riverside pub well away from the academic centre of Cambridge. Apart from the pint of beer in his hand, Aubrey has all the appearances of a guru, complete with a beard down to his navel, which he strokes thoughtfully as he launches into a well-rehearsed sketch of his rise to celebrity. His first theoretical papers, published in specialist gerontology journals, gave thought to the free-radical theory of ageing, which holds that ageing can be attributed to progressive damage to the body's cells inflicted by oxidants and other free radicals (molecules with unpaired electrons). De Grey proposed an intricate mechanism by which mutant mitochondrial DNA – the DNA located within what is effectively the engine-room of each cell – hampers the cells' ability to deal with free-radical attack. He expanded his thesis into a book, and was awarded a doctorate on the strength of it in 2000. He recognized then, though, that mitochondrial DNA was only one likely factor involved in ageing, not its single cause. His speculations became broader in scope and more polemical. He began to publish papers with provocative titles like 'An Engineer's Approach to the Development of Real Anti-Aging Medicine' and 'Is Human Aging Still Mysterious Enough to Be Left Only to Scientists?' He dared to discuss not only the arrest of ageing, but actually reversing it, and doing so 'within a matter of decades'. This bold promise launched him on the international conference circuit, where he thrived. His subsequent claim that we might soon expect to live to 1,000 years old was widely repeated in the media; he quickly tweaked it to suggest, excitingly, that the first person destined to live to 1,000 might be alive already.

And yet, Aubrey tells me, 'predictions about longevity are the least controversial things I am saying.' What really got him into hot water was that he itemized the breakthroughs that need to be made in order that we might live for another couple of decades, after which, he blithely claims, prolonging life still further becomes a whole lot easier. There are seven causes of death on the list, mostly to do with the body's replacement or not of cells and their contamination or damage by external factors, each and all of which will have to be successfully tackled if human life is to be significantly prolonged. The wish list made de Grey's project appear practical, and that worried orthodox biological researchers, who began to look as if they might be doing rather little about improving our life chances. 'I put people in a very conflicted position. They can't see a gap in my argument. They are terribly afraid I might be right,' Aubrey tells me.

The structures that de Grey has assembled around him further suggest a seriousness of purpose. He helped to set up the SENS Foundation – based in optimistic California, not in Cambridge – to pay for research into the prevention of ageing using charitable donations, and instituted the Methuselah Mouse Prize, an award for scientists who manage to extend the longevity of laboratory mice. Donors range from engineers to science fiction readers to fitness fanatics as well as those who give money in memory of lost loved ones.

Mainstream scientific research may have been lax. But I find Aubrey equally scornful of popular culture. Science fiction devoted to the idea of extended human lifespan draws surprising ire. 'It's clear that their speculation is for entertainment only,' he says. This sends the implicit message that death is acceptable. 'I find this absolutely tragic and appalling. Now that biotechnology has put us within serious striking distance, this whole issue of denial matters more. The origin of denial is pure terror. It's culturally universal. Only people who work in biogerontology themselves don't share this. They have other reasons for disliking me,' Aubrey adds teasingly.

De Grey saw the ugly side of the scientific establishment in 2005 when *Technology Review*, a respected magazine of the Massachusetts Institute of Technology, commissioned a profile of de Grey from Sherwin Nuland, who, as we have seen already, sides with what de Grey dismissively calls the 'supporters of ageing'. Opposing de Grey's visionary idealism, Nuland's tone was sage, magisterial – and fatalist;

he was happy to see human lifespan level off at a 'biologically prob-able maximum' of 120 years. The article was prefaced by an ill-judged editorial that amounted to an abusive ad hominem attack on de Grey. Yet such attacks merely play to de Grey's self-image of an embattled but righteous maverick. 'I'm at least at Gandhi stage three-and-a-half right now,' he says.

What unites de Grey with Voronoff and any other scientist who seeks to extend life is the fact that, at the level of cells, there is indeed immortality. Not all cells die. In particular, the germ cells exhibit what is known as 'biological immortality'. Why this should be so while other cells perish remains the focus of much research. Aware of this, the developmental biologist Lewis Wolpert – a man hardly shy of controversy in the defence of scientific rationalism – turns out to be surprisingly tolerant in his judgement of the immortalists. He does not believe they will succeed, but he does not, unlike the editors of *Technology Review*, who used the word on their front cover, dismiss them as 'nuts'. The germ cells – the egg and the sperm – do not age, and it is only the cells created thereafter in the developing embryo that are mortal. 'Potentially, then,' Wolpert conceded on BBC Radio in 2011, 'all causes of death are unnatural ones.'

Nothing irritates Aubrey de Grey more, I get the impression, than this recurrent rejoinder: yes, but what will we do with all the extra time? 'Erudite people become just embarrassing on this,' he says. Yet it's hardly a trivial issue. Human extension is pointless unless it is directed towards a goal. We use our technological extensions to go faster, to jump higher, to see the world in a different way. So why prolong life? What would it give us that we do not have already? I press the question. What about you personally? Aubrey struggles to think imaginatively about what he might do. 'It's completely crazy to be making decisions,' he splutters. 'More time to do what? I've absolutely no idea. And that's the point. My life has been relatively unpredictable so far, and that's great. It *is* extra time. But extra time is a side benefit. The whole thing is about *health*. My motives are humanitarian.

'If you're still biologically thirty at the age of eighty-five, golf will be losing its novelty,' he continues with a smile. 'So it's a chance to try other things. Serial careers and relationships are now much more the norm, so this simply extends a pattern.' Then he tries to spell it out with a joke: 'So many women, so much time.' It is clearly a catchphrase that pleases him, as I discover later that he has been using it for years. Yet it lays bare the big point he is missing. We live for ever by having children.

They may not reach conclusions that please de Grey, but I find that many of our stories explore extended longevity with considerable subtlety. Greatly aged characters have always been stock characters. In the Bible, Methuselah lived 969 years. I think this is an under-standable exaggeration. In former times when most people were dead by the age of thirty, quite a few would nevertheless have sur-vived to double or treble that age. It is not like that today, when we mostly die around the same age, and there is among us no equivalent representation of persons aged 150 or 200. Indeed, this statistical dif-ference might be taken as a clue that there is less scope to extend human life than de Grey thinks.

Methuselah's age is given in Genesis more or less as a matter of record. In more recent stories, however, superannuated characters are used to dramatize the moral dilemmas of ageing and mortality. They anticipate some of the social and economic issues that confront mod-ern gerontologists. For instance, the Struldbrugs in *Gulliver's Travels* grow aged even though they do not die, and must therefore be declared legally dead to prevent them from hoarding the wealth that could be enjoyed by younger generations.

But the story that most exactly captures the scenario envisioned by de Grey and his confreres of greatly – but perhaps not infinitely – extended lifespan, and extended prime of life rather than extended old age, is *The Makropulos Secret*, Karel Čapek's 1922 play later adapted as an opera by Leos Janáček. The titular secret is a formula developed in 1601 by one Hieronymus Makropulos for his patron, Emperor Rudolf II, which can prolong life by 300 years. Afraid that he might be poisoned, Rudolf demands that Makropulos try it out first on his sixteen-year-old daughter, Elina. The action of the play begins, how-ever, in 1922 in Prague, where a complex legal case has been dragging on for nearly a century. The glamorous singer Emilia Marty is a key witness, and turns out to have a strange familiarity with long-ago aspects of the case, in particular to do with a string of women, all with the initials E. M. Finally, Emilia tells her story – she is Elina, born in 1585, and has lived down the centuries since, periodically changing her name to avoid suspicion regarding her age, and leaving a trail of lovelorn admirers in her wake. Now, as the cynical Emilia Marta, tired of life, but afraid of death, she is the only one who knows where the formula is hidden, and she herself is in need of a top-up if she is to survive much longer. In the end, though, she chooses to forgo the chance to renew her life, and surrenders the formula. The protagonists and lawyers in the case all refuse it, and it is passed at last to the legal assistant's young daughter, who is an aspiring singer the same age as Elina was when she swallowed the potion. Without hesitation, she burns the formula, and Emilia / Elina finally expires at the splendid age of 337.

When Janáček saw the play, he was himself in the fruitful autumn of his career, rejuvenated by his passion for a much younger woman, Kamila Stösslová. He immediately set about adapting Čapek's clever comedy of ideas into a moving personal tragedy. 'We are happy because we know that our life isn't long,' he observed to Kamila. '[T]hat woman – the 337-year-old beauty – didn't have a heart any more.'

Čapek's theme is taken up by the philosopher Bernard Williams in an essay 'on the tedium of immortality'. Williams does not think it at all peculiar that E. M.'s existence had lost all meaning. 'The more one reflects to any realistic degree on the conditions of E. M.'s unending life, the less it seems a mere contingency that it froze up as it did,' he writes. For de Grey, this kind of talk is simple defeatism – and it's interesting to note, by the by, that Williams is careful himself not to be drawn on what might be an appropriate age for the freeze to begin. To go there would expose the weakness in his argument and the strength of the immortalists' case.

Of course, boredom is a poor response to the opportunities life offers at any age. E. M. has lived as several personas, and tired of each of them. She has tried the serial relationships that de Grey hankers after, and found even these wanting. Yet if we do have a list of things we are planning to do in our next century – make love to beautiful partners, write that novel, win Olympic gold, you can make your own list – we must ask ourselves why we are not doing each of those things now, while we definitely have the chance. The answers are different in each case, and some of them may surprise you.

Epilogue: Coming Home

While writing this book, I have been interrupted from time to time by news of public exhibitions with names like 'Human+' and 'Super-human', and even the publication of a book puzzlingly, though I suppose audaciously, entitled *Humanity 2.0*. I have learned that the terms 'posthuman' and 'transhuman' are certainly not confined to the genre of science fiction. I have read that our very flesh is at stake 'in our posthuman times', and that the 'boundaries between the human and nonhuman have been completely breached'. Another book carries the (optimistic? threatening?) subtitle: *When Humans Transcend Biology*.

But then I've read, too, about 'enhancement' and 'optimization' of the biological human body – if often with little sense of the direction in which improvement lies. I've seen how the emerging discipline of synthetic biology – using a set of technologies that enables functional biological devices to be fabricated from artificial starting materials – is encouraging not only life scientists but also engineers and designers to speculate in increasingly practical terms on the kind of changes we might make. 'The definition of "human" will expand,' states one not untypical manifesto. 'Our children's children will look nothing like us. And that will be by design.'

What I find startling in the rhetoric of both groups – the body tran-scenders and the body transformers – is their uncritical adoption of the language of consumer culture, with its implication that our own bod-ies are commodities to be ordered and chosen, bought and sold, and even taken back to the shop if we don't like them. This language is coloured in particular by the advertising jargon used to sell digital tech-nology. The Cartesian body-as-machine has been reinterpreted in the light both of modern medical science and the development of artificial intelligence to become the body-as-computer. We see raised before us a new body that we are invited to describe not in parts, but, as it were, in bits. The unspoken assumption behind this repositioning is that our species is due for, and deserving of, an upgrade.

Whereas the immortalists merely seek ways to live for longer or for ever in our own bodies, transhumanists disdain corporeal existence altogether, and wish to escape it. Their goal is to be able to 'upload' our minds to some grand ethereal network, and no longer to be dependent on flesh at all, or for that matter on the biosphere necessary to sustain it. (So far as I have been able to ascertain, the proponents of these fantasies are entirely men; much of the most thought-provoking philosophy of corporeality, on the other hand, comes from women, who seem more content (or resigned) to carry on living life within the bodies we have been given.)

None of this is new. The thought that the body is the prison of the soul goes back well beyond Descartes to Platonic philosophy. Present excitement about the disembodied mind, then, cannot be attributed solely to the technological moment in which we find ourselves. It speaks more of an acute discomfort and dissatisfaction with the body. Science reflects this discomfort with its relentless narrowing of focus on the smallest components of our biological existence. Artists are alive to it in a different way, exploiting our corporeal anxieties with a new turn to figurative art and hybrid projects to create tissue art and 'semi-living creatures'. At the same time, any public display of actual human bodies, for whatever purpose, and in whatever state of preparation or decay, arouses controversy.

The inference that the body is a mere nuisance moves us further than ever from any meaningful reconciliation of body and mind. Do we really aspire to escape from the body? If so, to where? To a better place, a place of safety, a place of order and regularity, a place of reliable and predictable performance? This dream is no extension of human life, but a denial of its real nature. It pretends that our minds are machines of our own brilliant devising: so enraptured are we with the computers we have invented that it seems we want to be more like them. It conveniently forgets that our minds are biological, too, and that they reside in, and depend upon, our bodies.

There is no escape. But this does not mean that we should regard as a prison what is in fact our home. It's quite a place.

References and Select Bibliography

General References

Aldersey-Williams, Hugh, Ken Arnold, Mick Gordon, Nikolaos Kotso-
poulos, James Peto and Chris Wilkinson, eds., *Identity and Identification*
(London: Black Dog Publishing, 2009).

Andrews, Lori, and Dorothy Nelkin, *Body Bazaar: The Market for Human
Tissue in the Biotechnology Age* (New York: Crown, 2001).

Aubrey, John, *Brief Lives*, ed. Oliver Lawson Dick (Harmondsworth: Penguin, 1972).

Barcan, Ruth, *Nudity: A Cultural Anatomy* (Oxford: Berg, 2004).

Blood, Sylvia K., *Body Work: The Social Construction of Women's Body Image*
(London: Routledge, 2005).

Butler, Judith, *Bodies That Matter* (London: Routledge 1993).

Bynum, Caroline W., *Metamorphosis and Identity* (New York: Zone, 2001).

Cartwright, Lisa, *Screening the Body: Tracing Medicine's Visual Culture* (Min-
neapolis: University of Minnesota Press, 1995).

Crooke, Helkiah, *Microcosmographia: A Description of the Body of Man Together
with the Controversies and Figures thereto belonging* (London: W. Iaggard, 1618).

Cunningham, Andrew, *The Anatomical Renaissance* (Aldershot: Ashgate,
1997).

Detsi-Diamanti, Zoe, Katerina Kitsi-Mitakou and Effie Yiannpoulou, *The
Future of Flesh: A Cultural Survey of the Body* (New York: Palgrave Macmillan, 2009).

Elias, Norbert, *The History of Manners* (New York: Pantheon, 1978).

Gage, John, *Colour and Culture* (London: Thames and Hudson, 1993).

Gallagher, Catherine, and Thomas Laqueur, eds., *The Making of the Modern
Body: Sexuality and Society in the Nineteenth Century* (Berkeley: University
of California Press, 1987).

Gilman, Sander L., *Making the Body Beautiful: A Cultural History of Aesthetic
Surgery* (Princeton: Princeton University Press, 1999).

Gould, Stephen Jay, *Eight Little Piggies* (London: Penguin, 1994).

Gould, Stephen Jay, *The Mismeasure of Man* (London: Penguin, 1997).

Gray, Henry, *Gray's Anatomy*, 1901 edn, ed. T. Pickering Pick (Philadelphia: Running Press, 1974).

Haraway, Donna, *Simians, Cyborgs and Women: The Reinvention of Nature* (London: Free Association Press, 1991).

Hillman, David, and Carla Mazzio, eds., *The Body in Parts: Fantasies of Corporeality in Early Modern Europe* (New York: Routledge, 1997).

Kemp, Martin, and Marina Wallace, *Spectacular Bodies: The Art and Science of the Human Body from Leonardo to Now* (Berkeley: University of California Press, 2000).

Kristeva, Julia, *The Powers of Horror: An Essay on Abjection*, trans. Leon Roudiez (New York: Columbia University Press, 1982).

Kuriyama, Shigehisa, *The Expressiveness of the Body and the Divergence of Greek and Chinese Medicine* (New York: Zone Books, 1999).

Kussi, Peter, ed., *Toward the Radical Center: A Karel Capek Reader* (North Haven, CT: Catbird Press, 1990).

Lock, Margaret, *Twice Dead: Organ Transplants and the Reinvention of Death* (Berkeley: University of California Press, 2002).

MacDonald, Helen, *Human Remains: Dissection and Its Histories* (New Haven, CT: Yale University Press, 2006).

Marieb, Elaine N., and Jon Mallatt, *Human Anatomy*, 3rd edn (San Francisco: Benjamin Cummings, 2001).

Martini, Frederic H., *Fundamentals of Anatomy and Physiology*, 6th edn (San Francisco: Benjamin Cummings, 2004).

Montaigne, Michel de, *The Complete Essays*, trans. M. A. Screech (London: Allen Lane, 1991).

Moore, Lisa Jean, and Mary Kosut, *The Body Reader: Essential Social and Cultural Readings* (New York: New York University Press, 2010).

Nuland, Sherwin B., *How We Die* (London: Chatto and Windus, 1994).

Nuland, Sherwin B., *How We Live: The Wisdom of the Body* (London: Chatto and Windus, 1997).

Onions, R. B., *The Origins of European Thought about the Body, the Mind, the Soul, the World, Time and Fate* (Cambridge: Cambridge University Press, 1951).

Orbach, Susie, *Bodies* (London: Profile, 2009).

Ovid, *Metamorphoses*, trans. David Raeburn (London: Penguin, 2004).

Petherbridge, Deanna, and L. J. Jordanova, *The Quick and the Dead: Artists and Anatomy* (London: Hayward Gallery, 1997).

Porter, Roy, *Flesh in the Age of Reason* (London: Allen Lane, 2003).

Rabelais, François, *Gargantua and Pantagruel*, trans. M. A. Screech (London: Penguin, 2006).

Richardson, Ruth, *Death, Dissection and the Destitute*, 2nd edn (Chicago: University of Chicago Press, 2000).

Rose, Nikolas, *Inventing Our Selves: Psychology, Power and Personhood* (Cambridge: Cambridge University Press, 1996).

Rousselet, Jean, ed., *Medicine in Art: A Cultural History* (New York: McGraw-Hill, 1967).

Saunders, Corinne, Ulrike Maude and Jane Macnaughton, eds., *The Body and the Arts* (Basingstoke: Macmillan, 2009).

Sawday, *The Body Emblazoned: Dissection and the Human Body in Renaissance Culture* (London: Routledge, 1995).

Schama, Simon, *Rembrandt's Eyes* (New York: Alfred A. Knopf, 1999).

Schupbach, William, 'The Paradox of Rembrandt's "Anatomy of Dr. Tulp"', *Medical History*, Supplement No. 2 (London: Wellcome Institute for the History of Medicine, 1982).

Sharp, Lesley A., *Strange Harvest: Organ Transplants, Denatured Bodies and the Transformed Self* (Berkeley: University of California Press, 2006).

Sharp, Lesley A., *Bodies, Commodities, and Biotechnologies* (New York: Columbia University Press, 2007).

Shelley, Mary, *Frankenstein, or The Modern Prometheus* (Harmondsworth: Penguin, 1994).

Shorter, Edward, *A History of Women's Bodies* (London: Allen Lane, 1983).

Stafford, Barbara Maria, *Body Criticism: Imaging the Unseen in Enlightenment Art and Medicine* (Cambridge, MA: MIT Press, 1991).

Sterne, Laurence, *The Life and Opinions of Tristram Shandy, Gentleman* (London: Penguin, 1997).

Sweet, Matthew, *Inventing the Victorians* (London: Faber and Faber, 2001).

Turner, Bryan S., *The Body and Society: Explorations in Social Theory*, 2nd edn (London: Sage, 1996).

Vesalius, Andreas, *De Humani Corporis Fabrica* (San Francisco: Norman Publishing, 1998).

Walters, Margaret, *The Male Nude: A New Perspective* (New York: Paddington Press, 1978).

Warner, Marina, *Monuments and Maidens: The Allegory of the Female Form* (London: Weidenfeld and Nicolson, 1985).

Welton, Donn, ed., *Body and Flesh: A Philosophical Reader* (Malden, MA: Blackwell, 1998).

Chapter References

Introduction

McCandless, David, *Information Is Beautiful* (London: Collins, 2010).

Prologue: The Anatomy Lesson

Broos, B., N. E. Middelkoop, P. Noble and J. Wadum, *Rembrandt under the Scalpel* (The Hague: Mauritshuis, 1998).

de Vries, A. B., M. Tóth-Ubbens and W. Froentjes, *Rembrandt in the Mauritshuis: An Interdisciplinary Study* (Alphen aan de Rijn: Sijthoff and Noordhoff, 1978).

Heckscher, W. S., *Rembrandt's Anatomy of Dr. Nicolaas Tulp: An Iconological Study* (New York: New York University Press, 1958).

Volkenandt, Claus, *Rembrandt: Anatomie eines Bildes* (Munich: Wilhelm Fink, 2004).

Mapping the Territory

Burke, Edmund, *A Philosophical Enquiry into the Origin of Our Ideas of the Sublime and Beautiful* (London: Routledge, 2008).

Carroll, Lewis, *Alice in Wonderland* (London: Macmillan, 1996).

Cartwright, Lisa, 'A Cultural Anatomy of the Visible Human Project', in Paula Treichler, Lisa Cartwright and Constance Penley, eds., *The Visible Woman: Imaging Technologies, Gender, and Science* (New York: New York University Press, 1998), pp. 21–43.

Epictetus, *The Discourses*, ed. and trans. W. A. Oldfather (London: Heinemann, 1926–8).

Galton, Francis, 'Pocket Registrator for Anthropological Purposes', *Nature*, vol. 22 (1880), 478.

Galton, Francis, *The Narrative of an Explorer in Tropical South Africa* (London: Ward, Lock, 1889).

Haldane, J. B. S., 'On Being the Right Size', *Harper's Magazine* (March 1926).

Jencks, Charles, *Le Corbusier and the Tragic View of Architecture* (Harmondsworth: Penguin, 1987).

Le Corbusier, *The Modulor* (Basel: Birkhaüser, 2004).

Leonardo da Vinci, *The Notebooks*, ed. Irma A. Richter (Oxford: Oxford University Press, 1980).

Spenser, Edmund, *The Faerie Queene* (Harmondsworth: Penguin, 1978).

Swift, Jonathan, *Gulliver's Travels* (London: Penguin, 2001).

Vitruvius, *Ten Books on Architecture*, trans. Ingrid D. Rowland (Cambridge: Cambridge University Press, 1999).

Waldby, Catherine, *The Visible Human Project* (London: Routledge, 2000).

Wittgenstein, Ludwig, *Philosophical Investigations* (Oxford: Blackwell, 1953/2001).

Flesh

Hillman, David, *Shakespeare's Entrails: Belief, Scepticism and the Interior of the Body* (Basingstoke: Palgrave Macmillan, 2007).

Johnson, Paul, *Elizabeth I: A Study in Power and Intellect* (London: Weidenfeld and Nicolson, 1974).

Kail, Aubrey C., *The Medical Mind of Shakespeare* (Sydney: MacLennan and Petty, 1986).

Levi-Navarro, Elena, *The Culture of Obesity in Early and Late Modernity: Body Image in Shakespeare, Jonson, Middleton, and Skelton* (New York: Palgrave Macmillan, 2008).

Morris, Richard, Heinrich Hupe and Max Kaluza, eds., *Cursor Mundi: A Northumbrian Poem of the XIVth Century* (London: Oxford University Press, 1962).

Pliny the Elder, *Natural History: A Selection*, ed. and trans. John F. Healy (London: Penguin, 2004).

Swami, Viren, Maggie Gray and Adrian Furnham, 'The Female Nude in

Rubens: Disconfirmatory Evidence of the Waist-to-Hip Ratio Hypothesis of Female Attractiveness', *Imagination, Cognition and Personality*, vol. 26 (2006–7), 139–47.

Bones

Bogin, Barry, *The Growth of Humanity* (New York: Wiley, 2001).

Brecher, Ruth, and Edward M. Brecher, *The Rays: A History of Radiology in the United States and Canada* (Baltimore: Williams and Wilkins, 1969).

Gordon, J. E., *Structures, or Why Things Don't Fall Down* (London: Penguin, 1991).

Mann, Thomas, *The Magic Mountain*, trans. H. T. Lowe-Porter (London: Secker and Warburg, 1946).

McGrath, Roberta, *Seeing Her Sex: Medical Archives and the Female Body* (Manchester: Manchester University Press, 2002).

Paley, W., *Natural Theology, or Evidences of the Existence and Attributes of the Deity*, 12th edn (London: J. Faulder, 1809).

Shipman, Pat, Alan Walker and David Bichell, *The Human Skeleton* (Cambridge, MA: Harvard University Press, 1985).

Stokes, Ian A. F., *Mechanical Factors and the Skeleton* (London: John Libbey, 1981).

Carving Up the Territory

Atlas, Michel C., 'Ethics and Access to Teaching Materials in the Medical Library: The Case of the Pernkopf Atlas', *Bulletin of the Medical Libraries Association*, vol. 89, no. 1 (2001), 51–8.

Darwin, Charles, *The Origin of Species* (London: Penguin, 1985).

Dodd, Philip, *The Reverend Guppy's Aquarium* (London: Random House, 2007).

Hayes, Bill, *The Anatomist: A True Story of Gray's Anatomy* (New York: Ballantine, 2008).

Hunter, John, *The Natural History of the Human Teeth* (London: J. Johnson, 1778).

Hunter, William, *The Anatomy of the Human Gravid Uterus* (Birmingham: John Baskerville, 1774).

Knox, Robert, *A Manual of Artistic Anatomy* (London: H. Renshaw, 1852).

Lawrence, Susan C., and Kate Bendixen, 'His and Hers: Male and Female Anatomy in Anatomy Texts for US Medical Students', *Social Science and Medicine*, vol. 35 (October 1992), 925–33.

McCulloch, N. A., D. Russel and S. W. McDonald, 'William Hunter's Gravid Uterus: The Specimens and Plates', *Clinical Anatomy*, vol. 15, no. 4 (2002), 253–62.

O'Connell, Helen E., Kalavampara V. Sanjeevan and John M. Huston, 'Anatomy of the Clitoris', *Journal of Urology*, vol. 174 (2005), 1189–95.

O'Malley, C. D., *Andreas Vesalius of Brussels 1514–1564* (Berkeley: University of California Press, 1964).

Pernkopf, Eduard, *Atlas der topographischen und angewandten Anatomie des Menschen* (Munich: Urban und Schwarzenberg, 1963); translated as *Atlas of Topographical and Applied Human Anatomy*, ed. Helmut Ferner, trans. Harry Monsen (Philadelphia: W. B. Saunders Company, 1963–4).

Rae, Isobel, *Knox the Anatomist* (Edinburgh: Oliver and Boyd, 1964).

Richardson, Ruth, *The Making of Mr Gray's Anatomy* (Oxford: Oxford University Press, 2008).

Shelton, Don, 'The Emperor's New Clothes', *Journal of the Royal Society of Medicine*, vol. 105 (2010), 46–50.

Sque, Magi, and Sheila Payne, *Organ and Tissue Donation: An Evidence Base for Practice* (Maidenhead: Open University Press, 2007).

The Head

Cooper, Wendy, *Hair: Sex, Society, Symbolism* (London: Aldus Books, 1971).

Courtenay-Smith, Natasha, 'Why I'll Never Wear Hair Extensions Again, by Pop Star Jamelia', *Daily Mail* (18 July 2008).

de Beauvoir, Simone, *Brigitte Bardot and the Lolita Syndrome* (New York: Arno Press, 1972).

Doddi, N. M., and R. Eccles, 'The Role of Anthropometric Measurements in Nasal Surgery and Research: A Systematic Review', *Clinical Otolaryngology*, vol. 35 (2010), 277–83.

Gogol, Nikolai, *Plays and Petersburg Tales*, trans. Christopher English (Oxford: Oxford University Press, 1995).

Kershaw, Alister, *A History of the Guillotine* (London: J. Calder, 1958).

Meijer, Miriam Claude, *Race and Aesthetics in the Anthropology of Petrus Camper (1722–1789)* (Amsterdam: Rodopi, 1999).

Stora, Elie, *Un Médecin au XVIIe Siècle: François Bernier (1620–1688)* (Paris: Librarie Medicale Marcel Vigne, 1937).

The Face

Bruce, Vicki, and Andy Young, *In the Eye of the Beholder: The Science of Face-Perception* (Oxford: Oxford University Press, 1998).

Bull, Ray, and Nichola Rumsey, *The Social Psychology of Facial Appearance* (New York: Springer Verlag, 1988).

Davis, Natalie Zemon, *The Return of Martin Guerre* (Cambridge, MA: Harvard University Press, 1983).

Galton, Francis, [composite photographs], Galton Papers 158/2 (UCL Special Collections).

Galton, Francis, 'Generic Images', *Proceedings of the Royal Institution*, vol. 9 (1879), 161–70.

Galton, Francis, *Inquiries into Human Faculty and Its Development* (London: Macmillan, 1883).

Galton, Francis, *Memories of My Life* (London: Methuen, 1908).

Grann, David, 'The Chameleon', *New Yorker* (11 August 2008).

Hume, David, *A Treatise of Human Nature*, ed. L. A. Selby-Bigge and P. H. Nidditch (Oxford: Clarendon Press, 1978).

Langlois, Judith, and Lori Roggman, 'Attractive Faces Are Only Average', *Psychological Science*, vol. 1 (1990), 115–21.

Lavater, Johann Caspar, *Essays on Physiognomy: For the Promotion of the Knowledge and the Love of Mankind*, trans. Thomas Holcroft (London: G. G. J. and J. Robinson, 1789).

Roe, John O., 'A Classic Reprint: The Deformity Termed "Pug-Nose" and Its Correction by a Simple Operation', *Aesthetic Plastic Surgery*, vol. 10 (1986), 89–91.

Royal College of Surgeons, *Facial Transplantation Working Party Report*, 2nd edn (London: Royal College of Surgeons, 2006).

The Brain

Aldersey-Williams, Hugh, 'Sharpest Look Yet inside the Body', *Popular Science* (June 1988).

Anderson, Britt, and Thomas Harvey, 'Alterations in Cortical Thickness and Neuronal Density in the Frontal Cortex of Albert Einstein', *Neuroscience Letters*, vol. 210 (1996), 161–4.

Combe, George, 'On the Cerebral Development and Moral and Intellectual Character of Raphael Sanzio d'Urbino', *Phrenological Journal and Magazine of Moral Science*, vol. 19 (1846), 42–55.

Diamond, Marian C., Arnold B. Scheibel, Greer M. Murphy Jr and Thomas Harvey, 'On the Brain of a Scientist: Albert Einstein', *Experimental Neurology*, vol. 88 (1985) 198–204.

Gall, Franz Joseph, *On the Functions of the Brain and of Each of Its Parts* (Boston: Marsh, Capen and Lyon, 1835).

Greene, Joshua D., Leigh E. Nystrom, Andrew D. Engell, John M. Darley and Jonathan D. Cohen, 'The Neural Bases of Cognitive Conflict and Control in Moral Judgment', *Neuron*, vol. 44 (2004), 389–400.

Hare, Todd A., Colin F. Camerer and Antonio Rangel, 'Self-Control in Decision-Making Involves Modulation of the vmPFC Valuation System', *Science*, vol. 324 (2009), 646–8.

Highfield, Roger, and Paul Carter, *The Private Lives of Albert Einstein* (London: Faber and Faber, 1993).

Kevles, Bettyann Holtzmann, *Naked to the Bone: Medical Imaging in the Twentieth Century* (New Brunswick, NJ: Rutgers University Press, 1997).

Limb, C. J., and A. R. Braun, 'Neural Substrates of Spontaneous Musical Performance: An fMRI Study of Jazz Improvisation', *PLoS ONE*, vol. 3, no. 2 (2008), e1679.

'London Lyrics', *New Monthly Magazine and Literary Journal*, vol. 5 (1823), 428.

Mitchell, Jason P., C. Neil Macrae and Mahzarin R. Banaji, 'Dissociable Medial Prefrontal Contributions to Judgments of Similar and Dissimilar Others', *Neuron*, vol. 50 (2006), 655–63.

'On the Heads and Intellectual Qualities of Sir Isaac Newton and Lord Bacon', *Phrenological Journal and Magazine of Moral Science*, vol. 18 (1845), 153–6.

Penfield, Wilder, and Theodore Rasmussen, *The Cerebral Cortex of Man: A Clinical Study of Localization of Function* (New York: Macmillan, 1950).

Reeve, Henry, journal of travels, 1805–6 (unpublished: Wellcome Library MS.5430).

Schott, G. D., 'Penfield's Homunculus: A Note on Cerebral Cartography', *Journal of Neurology, Neurosurgery, and Psychiatry*, vol. 56 (1993), 329–33.

Simpson, James, 'Observations on Some Recent Objections to Phrenology, Founded on a Part of the Cerebral Development of Voltaire', *Phrenological Journal and Miscellany*, vol. 3 (1825), 564–78.

Tomlinson, Stephen, *Head Masters: Phrenology, Secular Education, and Nineteenth-Century Social Thought* (Tuscaloosa, AL: University of Alabama Press, 2005).

Walton, Mark E., Joseph T. Devlin and Matthew F. S. Rushworth, 'Interactions between Decision Making and Performance Monitoring within Prefrontal Cortex', *Nature Neuroscience*, vol. 7 (2004), 1259–65.

Wells, Samuel R., *The Illustrated Annual of Phrenology and Physiognomy* (New York: Fowler and Wells, 1869).

Witelson, Sandra F., Debra L. Kigar and Thomas Harvey, 'The Exceptional Brain of Albert Einstein', *Lancet*, vol. 353 (1999), 2149–53.

The Heart

Alberti, Fay Bound, *Matters of the Heart: History, Medicine and Emotion* (Oxford: Oxford University Press, 2010).

Armour, J. Andrew, 'The Little Brain on the Heart', *Cleveland Clinical Journal of Medicine*, vol. 74 (2007), S48–S51.

Caplan, Arthur L., *If I Were a Rich Man Could I Buy a Pancreas?* (Bloomington: Indiana University Press, 1992).

Erickson, Robert A., *The Language of the Heart, 1600–1750* (Philadelphia: University of Pennsylvania Press, 1997).

Fox, Renée C., and Judith P. Swazey, *Spare Parts: Organ Replacement in American Society* (New York: Oxford University Press, 1992).

Kresh, J. Yasha, and J. Andrew Armour, 'The Heart as a Self-Regulating System: Integration of Homeodynamic Mechanisms', *Technology and Health Care*, vol. 5 (1997), 159–69.

Nabokov, Vladimir, *Bend Sinister* (London: Weidenfeld and Nicolson, 1960).

Peto, James, ed., *The Heart* (New Haven, CT: Yale University Press, 2007).

Siegel, Jason T., and Eusebio M. Alvaro, eds., *Understanding Organ Donation* (Chichester: Wiley-Blackwell, 2010).

Thompson, D'Arcy, *On Growth and Form* (Cambridge: Cambridge University Press, 1992).

Titmuss, Richard M., *The Gift Relationship: From Human Blood to Social Policy* (London: George Allen and Unwin, 1970).

Vinken, Pierre, *The Shape of the Heart* (Amsterdam: Elsevier, 2000).

Young, Louisa, *The Book of the Heart* (London: HarperCollins, 2002).

Blood

Arikha, Noga, *Passions and Tempers: A History of the Humours* (New York: Ecco, 2007).

Behrmann, Jason, and Vardit Ravitsky, 'Do Canadian Researchers Have "Blood on their Hands"?', *Canadian Medical Association Journal*, vol. 183 (2011), 1112.

Harvey, William, *Exercitatio Anatomica de Motu Cordis et Sanguinis in Animalibus* (Frankfurt: Sumptibus Guilielmi Fitzeri, 1628).

NHS Blood and Transplant, *Strategic Plan 2011–14* (Watford: NHSBT, 2011).

Piliavin, Jane Allyn, and Peter L. Callero, *Giving Blood: The Development of an Altruistic Identity* (Baltimore: Johns Hopkins University Press, 1991).

The Ear

Bronkhurst, Hans, *Vincent van Gogh* (London: Weidenfeld and Nicolson, 1990).

Carlyle, Thomas, *History of Friedrich II of Prussia, Called Frederick the Great* (London: Chapman and Hall, 1858–65).

Cohen, Ben, 'A Tale of Two Ears', *Journal of the Royal Society of Medicine*, vol. 96, no. 6 (June 2003), 305–6.

Hughes, Robert, *Nothing If Not Critical* (London: Collins Harvill, 1990).

Kaufmann, Hans, and Rita Wildegans, *Van Goghs Ohr: Paul Gauguin und der Pakt des Schweigens* (Berlin: Osborg, 2008).

Pirsig, Wolfgang, 'The Auricle in Visual Art', *Facial Plastic Surgery*, vol. 20, no. 4 (2004), 251–66.

Pirsig, Wolfgang, and Jacques Willemots, eds., *Ear, Nose and Throat in Culture* (Oostende: G. Schmidt, 2001).

Walpole, Horace, *Anecdotes of Painting in England* (London: J. Dodsley, 1782).

Weber, Bruce, 'J. Paul Getty III Dies; Had Ear Cut Off by Captors', *New York Times* (7 February 2011).

The Eye

Armel, K. C., and V. S. Ramachandran, 'Acquired Synesthesia in Retinitis Pigmentosa', *Neurocase*, vol. 5 (1999), 293–6.

Bach-y-Rita, Paul, Carter C. Collins, Frank A. Saunders, Benjamin White and Lawrence Scadden, 'Vision Substitution by Tactile Image Projection, *Nature*, vol. 221 (1969), 963–4.

Botvinick, M., and J. Cohen, 'Rubber Hands "Feel" Touch that the Eye Sees', *Nature*, vol. 391 (1998), 756.

Collignon, Olivier, 'Functional Specialization for Auditory-Spatial Processing in the Occipital Cortex of Congenitally Blind Humans', *Proceedings of the National Academy of Sciences*, vol. 108 (2011), 4435–40.

Descartes, René, *La Dioptrique* (Leiden, 1637).

Eiberg, Hans, et al., 'Blue Eye Color in Humans May Be Caused by a Perfectly Associated Founder Mutation in a Regulatory Element Located Within the *HERC2* Gene Inhibiting *OCA2* Expression', *Human Genetics*, vol. 123 (2008), 1777–87.

Galton, Francis, 'Family Likeness in Eye-Colour', *Nature*, vol. 34 (1886), 137; *Proceedings of the Royal Society*, vol. 40 (1886), 402–16.

Gaukroger, Stephen, John Schuster and John Sutton, *Descartes' Natural Philosophy* (London: Routledge, 2000).

Gregory, R. L., and J. G. Wallace, 'Recovery from Early Blindness: A Case Study', *Experimental Psychology Society Monograph*, No. 2 (1963).

Harrison, John E., *Synaesthesia: The Strangest Thing* (New York: Oxford University Press, 2001).

Ings, Simon, *The Eye: A Natural History* (London: Bloomsbury, 2007).

Nabokov, Vladimir, *Speak, Memory: An Autobiography Revisited* (London: Weidenfeld and Nicolson, 1967).

Ogden, Daryl, *The Language of the Eyes: Science, Sexuality and Female Vision in*

English Literature and Culture, 1690–1927 (New York: State University of New York Press, 2005).

Thompson, William I., *Coming into Being: Artifacts and Texts in the Evolution of Consciousness* (New York: St Martin's Press, 1996).

The Stomach

Arens, William, *The Man-Eating Myth: Anthropology and Anthropophagy* (New York: Oxford University Press, 1979).

Brillat-Savarin, Jean Anthelme, *The Physiology of Taste*, trans. M. F. K. Fisher (New York: Alfred A. Knopf, 2009).

Buckland, William, 'Agriculture', *Quarterly Review*, vol. 73 (1844), 477–509.

Burgess, G. H. O., *The Curious World of Frank Buckland* (London: John Baker, 1967).

Hare, A. J. C., *The Story of My Life*, vol. 5 (London: George Allen, 1900).

Owen, Richard, *The Life of Richard Owen* (London: John Murray, 1894–5).

Tiffin, Helen, 'Pigs, People and Pigoons', in Laurence Simmons and Philip Armstrong, eds., *Knowing Animals* (Leiden: Brill, 2007), pp. 244–65.

The Hand

Bulwer, John, *Chirologia and Chironomia* (London, 1644).

Carroll, Lewis, *Through the Looking-glass and What Alice Found There* (London: Macmillan, 1996).

Coren, Stanley, 'Left-Handedness and Accident-Related Injury Risk', *American Journal of Public Health*, vol. 79, no. 8 (August 1989), 1040–41.

Hepper, Peter, 'Handedness in the Human Foetus', *Neuropsychologia*, vol. 29, no. 11 (1991), 1107–11.

Hepper, Peter G., Deborah L. Wells and Catherine Lynch, 'Prenatal Thumb Sucking Is Related to Postnatal Handedness', *Neuropsychologia*, vol. 43, no. 3 (2005), 313–15.

McManus, Chris, *Right Hand, Left Hand* (London: Weidenfeld and Nicolson, 2002).

Napier, John, *Hands* (London: George Allen and Unwin, 1980).

Rahman, A. A., et al., 'Hand Pattern Indicates Prostate Cancer Risk', *British Journal of Cancer*, vol. 104 (2011), 175–7.

Smith, Richard Langham, programme note to CD 452 448-2 (London: Decca, 1996).

Tallis, Raymond, *Michelangelo's Finger* (London: Atlantic Books, 2010).

Tallis, Raymond, *The Hand: A Philosophical Enquiry into Human Being* (Edinburgh: Edinburgh University Press, 2003).

Trumble, Angus, *The Finger: A Handbook* (New Haven, CT: Yale University Press, 2010).

Weyl, Hermann, *Symmetry* (Princeton, NJ: Princeton University Press, 1952).

Wilson, Glenn D., 'Fingers to Feminism: The Rise of 2D:4D', *Quarterly Review*, vol. 4 (2010), 25–32.

Wolff, Charlotte, *The Human Hand* (London: Methuen, 1942).

Wolpert, Lewis, 'Development of the Asymmetric Human', *Biological Review*, vol. 13, suppl. 2 (2003), 97–103.

The Sex

Ashley, April, with Douglas Thompson, *The First Lady* (London: John Blake, 2006).

Barthes, Roland, *Mythologies*, trans. Annette Lavers (New York: Hill and Wang, 1972).

Clark, Kenneth, *The Nude: A Study of Ideal Art* (London: John Murray, 1956).

Cook, Edward Tyas, *The Life of John Ruskin* (London: G. Allen, 1911).

Davidson, Keay, *Carl Sagan: A Life* (New York: Wiley, 1999).

Harper, Catherine, *Intersex* (Oxford: Berg, 2007).

Howard, Seymour, 'Fig Leaf, Pudica, Nudity, and Other Revealing Concealments', *American Imago*, vol. 43 (1986), 289–93.

Johnson, Olive Skene, *The Sexual Spectrum: Why We're All Different* (Vancouver: Raincoast Books, 2004).

Sagan, Carl, *The Cosmic Connection: An Extraterrestrial Perspective* (Garden City, NY: Anchor Press, 1973).

Stainton Rogers, Wendy, and Rex Stainton Rogers, *The Psychology of Gender and Sexuality* (Buckingham: Open University Press, 2001).

The Foot

Defoe, Daniel, *Robinson Crusoe* (London: Penguin, 2001).

Hume, David, *Enquiries Concerning Human Understanding and Concerning the Principles of Morals*, ed. L. A. Selby-Bigge and P. H. Nidditch (Oxford: Clarendon Press, 1975).

Mark, Darren F., Silvia Gonzales, David Huddart and Harald Böhnel, 'Dating of the Valsequillo Volcanic Deposits: Resolution of an Ongoing Archaeological Controversy in Central Mexico', *Journal of Human Evolution*, vol. 58 (2010), 441–5.

McAllister, Peter, *Manthropology: The Science of the Inadequate Modern Male* (New York: St Martin's Press, 2010).

Webb, Steve, 'Further Research of the Willandra Lakes Fossil Footprint Site, Southeastern Australia', *Journal of Human Evolution*, vol. 52 (2007), 711–15.

Webb, Steve, Matthew L. Cupper and Richard Robins, 'Pleistocene Human Footprints from the Willandra Lakes, Southeastern Australia', *Journal of Human Evolution*, vol. 50 (2006), 405–13.

The Skin

'A Lover of Physick and Surgery', *An Essay Concerning the Infinite Wisdom of God, Manifested in the Contrivance and Structure of the Skin* (London: Joseph Marshall, 1724).

Ableman, Paul, *Anatomy of Nakedness* (London: Orbis, 1982).

Beales, Peter, *Classic Roses* (London: Harvill, 2004).

Caplan, Jane, ed., *Written on the Body: The Tattoo in European and American History* (London: Reaktion, 2000).

Connor, Steven, *The Book of Skin* (London: Reaktion, 2004).

Darwin, Charles, *The Expression of the Emotions in Man and Animals* (London: John Murray, 1872).

Hoffman, Banesh, *Einstein: Creator and Rebel* (New York: Viking, 1972).

Lutyens, Mary, *Millais and the Ruskins* (London: John Murray, 1967).

Thompson, Peter, 'Margaret Thatcher: A New Illusion', *Perception*, vol. 9 (1980), 483–4.

Thomson, Arthur, *Handbook of Anatomy for Art Students* (London: Macmillan, 1896).

Woolner, Amy, *Thomas Woolner, R. A., Sculptor and Poet: His Life in Letters* (London: Chapman and Hall, 1917).

Extending the Territory

de Grey, Aubrey D. N. J., et al., 'Is Human Aging Still Mysterious Enough to Be Left Only to Scientists?', *BioEssays*, vol. 24 (2002), 667–76.

Deschamps, J.-Y., Françoise A. Roux, Pierre Saï and Edouard Gouin, 'History of Xenotransplantation', *Xenotransplantation*, vol. 12, no. 2 (2005), 91–109.

Lewis, C. S., *The Screwtape Letters* (London: Collins, 1979).

McLuhan, Marshall, *Understanding Media* (New York: McGraw-Hill, 1964).

Newrick, P. G., E. Affie and R. J. M. Corrall, 'Relationship Between Longevity and Lifeline: A Manual Study of 100 Patients', *Journal of the Royal Society of Medicine*, vol. 83 (August 1990), 499–501.

Nuland, Sherwin B., 'Do You Want to Live Forever?', *Technology Review*, vol. 108 (February 2005), 36–45.

Voronoff, Serge, *Quarante-trois greffes du singe à l'homme* (Paris: Doin, 1924).

Voronoff, Serge, *The Study of Old Age and My Method of Rejuvenation* (London: Gill Publishing, 1926).

Warner, Huber, et al., 'Science Fact and the SENS Agenda', *EMBO Reports*, vol. 6 (2005), 1006–8.

Williams, Bernard, *Problems of the Self* (Cambridge: Cambridge University Press, 1973).

Wilmoth, J. R., 'Demography of Longevity: Past, Present, and Future Trends', *Experimental Demography*, vol. 35 (2000), 1111–29.

Zemanová, Mirka, *Janáček* (Boston: Northeastern University Press, 2002).

Epilogue: Coming Home

Barringer, David, 'Self Created', *RSA Journal* (Winter 2011), 50.

Fuller, Steve, *Humanity 2.0* (Basingstoke: Palgrave Macmillan, 2011).

Index

Page references in *italic* indicate illustrations.